Practical Engineering Application
in Electrical Engineering Studies

Practical Engineering Application in Electrical Engineering Studies

Prepared by

Dr. Mohamed Said Khorsheed

Graduate of the Universities of The United
States of America, Formerly Certified Registered
Professional Engineer by Examination of the
Board of American Professional Engineers

To order additional copies of this book, contact:
Xlibris
1-888-795-4274
www.Xlibris.com
Orders@Xlibris.com
752894

CONTENTS

ABOUT THE AUTHOR

THE AUTHOR, AFTER receiving his high school education, attended the University of Texas in Austin, Texas, where he obtained his degree as an electrical engineer. He then continued his postgraduate engineering studies at the University of Santa Clara, and later the University of California in California.

The author gained his professional engineering experience by working for engineering firms in the United States, Europe, and the Middle East. He held such positions as electrical engineer, principal engineer, and nuclear electrical lead engineer. The author worked on the design and construction of major conventional and nuclear power generation plants' auxiliary power distribution systems, including the design and construction of industrial and petrochemical facilities in the United States, Europe, and the Middle East for over forty-five years of his professional life.

The following are firms where the author worked:

- Sargent & Lundy Engineers in the city of Chicago, Illinois, USA, a firm that specializes in the design and engineering of conventional and nuclear power plants.
- Bechtel Engineers & Constructors in the city of San Francisco, California, USA, a firm that specializes in the design, engineering, and construction of conventional and nuclear power plants, in addition to the design and construction of industrial and petrochemical plants.
- Brown & Root Engineers & Constructors in the city of Houston, Texas, USA, a firm that specializes in the design, engineering, and construction of conventional and nuclear

power plants, in addition to design and construction of industrial and petrochemical plants.

- Flour Engineers and Constructors in the city of San Mateo, California, USA, a firm that specializes in the design, engineering, construction, and construction management of industrial and petrochemical plants.

- American Arabian Oil Company in the city of Houston, Texas, USA, and in the Kingdom of Saudi Arabia in Saudi Arabia (ARAMCO), a major oil company owned by several oil firms in the United States and the Saudi government in Saudi Arabia, which later became a company solely owned by the Saudi government.

- Major Engineering Consultant and Technical Studies, a firm owned by the Syrian government in Syria, which specializes in the consultancy of several types of design, engineering, and construction management of projects owned by the local government in Syria.

- The author is a former licensed professional engineer by (written and oral) examinations sponsored by The American Board of Professional Engineers in the states of Texas, Illinois, and the California. He was given the professional authority to lawfully sign engineering and design documents to be effectively recognized by the local lawful authority in each of the aforementioned states.

- The author has written and published several technical engineering documents and books to be used as references for the engineering profession, among which are the books titled *Nuclear Efforts Studies, a Science and Learning Necessity*, and *Practical Engineering Application in Electrical Engineering Studies*.

- The author has written several technical articles educating the public media about the pros and cons of nuclear energy use alerting the media and society about their safety and the safety of their animals and nature when exposed to wrong use of the nuclear materials.

The book further described the type of the effect of nuclear radiation on those soldiers who served in the armed forces when those soldiers returned back home and got married.

The soldiers sequential exposure to the effect of the so-called secondary nuclear radiation emission had later produced negative harm in babies' chromosomes of their biological structure in the biological cells to produce the so-called crazy cells, which failed to carry the normal human biological heritage from father/mother to the soldiers' new babies.

The author's nuclear technical book has addressed the pros and cons with respect to the use of nuclear energy. The book also aimed at educating, in general, the media about the safe use of nuclear energy when such use has adhered to the application of the requirement of the Atom Energy Commission standards and regulations, and wherever such a commission has been established throughout the world.

PREFACE

PEOPLE, IN GENERAL, start early in age to build and strengthen their inspiration of building the type of social personality they desire; such a personality can be to become an engineer, a medical doctor, a lawyer, or even a teacher.

Whatever the case may be, the esteem of the educated individual shall be aimed in favor of serving the community in honesty, and to apply his careful know-how and best effort to serve and render such noble service to the community where he was meant to serve.

The engineering and design of any project depend on the cooperative efforts among all engineering disciplines who will work on a specific project, and without such safe and reliable effort, the project on hand will not be complete.

Developed nations place great emphasis on the safety of the nuclear power generation plants or conventional power generation projects design and engineering based on their compliance with the applicable engineering standards and regulations, and these restrictions are a must for the safety of the environment and the safety of the people who will be part of operating the completely constructed projects.

The book has referred to those addressed standards where applicable and insisted on the application of those standards and regulations, which the engineer shall be aware of and get used to in his effort to design and engineer projects to meet all of their requirement, which will ensure human safety requirements, including the safety of the environment which we live in.

In the following pages of this book, we shall talk in a comprehensive but not very detailed manner about the application of disciplines of the engineering profession in general and the application of electrical engineering in more detail. However, the specialized engineer must have the required academic background that he prepared himself for during his academic study. Such study shall include but is not limited to the study of mathematics, physics, chemistry, graphics, engineering economics, and the ability to master the language of those courses.

The prospective engineer studies of his engineering branch (major study of his/her curriculum) will include the know-how to prepare required engineering reports, graphs, and other comprehensive requirements, which will help him to execute a safe, engineered, and designed project that meets all applicable safety regulations and standards. The specialized engineer is required also (in some universities) to take additional courses in social studies, such as history, government, foreign languages, and other elective courses related to his major study, in addition to technical writing courses, which will help him prepare technical reports and technical specification during the practice of his/hers specific studied discipline.

It is imperative to include the practical laboratory studies associated with some courses of the engineer specialty during the academic year, and the preparation of technical reports that are related to the course of the study.

The laboratory studies constitute a major step in understanding the nature of the engineer major study, when laboratory experiments are pursued further in the future, which may lead the engineer to a new discovered application in his/hers future life.

Fortunately enough, I happened to fall in that category during the senior year of my college study, when a group of my friends and I prepared a laboratory experiment as a graduation project requirement, which dealt with a device called X and Y Electronic Plotters, which became later the invention foundation of the present electronic computers.

It is important in this book to address in general the requirements of the variety of engineering branches application at the time when addressing in more details the electrical engineering branch practical application requirements.

The prospective electrical engineer, when he underwent the period of his undergraduate study in college has learned that there are three distinct divisions in the electrical engineering study, namely the power division, electronics division, and communication division. These divisions interact among each other as much as the studied project requires to, in addition to the requirement of the mechanical engineering branch, or the civil engineering branch during the addressed project design work.

It is worth mentioning here that the application of power engineering design in any prospective project design may not play the major role, and yet it is very important to address its requirement as effectively as the major division application requirement (mechanical project or other project) in the addressed project design requirement.

The electrical energy has also played an effective role to support the operation of many industrial facilities, such as petrochemical plants and other manufacturing plants, such as electronics equipment production facilities and other facilities.

It is imperative to start this research by pointing to the fact how electrical energy is produced since it is the subject of this study; not only that but also to introduce the prospective engineer to the several types of energy-generating facilities, and the application of such electrical energy in powering many industrial facilities as listed in this discussion.

The important matter in the practical engineering application in any electrical engineering studies is first to start with the preparation of a special study called project feasibility study for the prospective project in accordance with the project owner (scope of work). This task details will be addressed in this book.

It is a procedure which will be discussed in this book in a comprehensive manner in order to aid the young engineer to prepare such task when it is required.

The electrical practical engineering study for any facility starts in general with the study of power delivery to the facility and power distribution to the electrical load throughout the facility as required with respect to the nature of the mechanical load. It later studies the requirement of locating the power supply centers near as possible to the mechanical loads. Then the study continues to locate the required cables raceway among the power supply centers and the mechanical load, taken in consideration the requirement of the local codes and regulations where the prospective project is to be built.

The entire effort of this partial study shall be coordinated with the other branches of engineering working on the design and engineering of the prospective project.

The prospective project is usually managed by a project manager who is mostly either a mechanical engineer or chemical engineer depending on the nature of the prospective project; hence, the requirement of controlling any equipment related to the nature of the project will be dictated by that special control engineer to submit an approved logic control diagram to the specialized electrical engineer to design the schematic control diagram for the equipment in question.

The coordinated design effort is an elaborate method that requires all branches of engineering involved in the engineering study of the prospective project.

The electrical engineering task continue to the following work procedure of doing engineering calculation and other electrical engineering tasks during the process of designing a complete design and engineering work, which will be the subject of this book and will be followed step-by-step depending on the nature of the project.

CHAPTER 1

Project Engineering Feasibility Study

PROJECT ENGINEERING FEASIBILITY study is considered the first step in any project engineering study for projects sized medium and larger. The study covers the review of the land where the project is going to be built on for the first time. Other projects, which are classified to be expansion projects, may not need a project engineering feasibility study unless the prospective project owner asks for such a study.

The purpose of the feasibility study in general is required when the prospective owner needs to arrange for financial cost, whether it is to issue new bonds in the name of his company or to borrow funds from financial institutes in order to support the cost of the new project.

The study in question usually helps the owner in general to take a decision whether he should pursue the execution of the new project or not when he compares the general cost of the project with his financial standing without any future burden. Not only that but he should be sure that the future project will bring adequate profit on top of the cost of the new project based on a certain economical study that his accounting will perform.

It should be emphasized here that the owner of the prospective project generally takes his final decision based on the actual financial ability to sponsor such a project, and that needs the company's board of directors' approval who has studied the approved feasibility study of the project on hand and reviewed the economic study of the project to his satisfaction.

It is well-known that some of the engineering firms who prepare engineering feasibility studies are also capable of preparing economic studies for the project on hand, which will be based on the project owner requirements and on the owner accomplished economic data, or sometimes based on the selected financial institute's requirements, which may finance the cost of the project on hand.

It is also worth mentioning here that the cost of the prospective project should include the cost of the land where the project to be built on, the cost of all required material, the cost of all needed construction works, and the cost of the project detailed engineering works.

Usually it is recommended to add a certain percentage of the overall cost to the estimated cost to take care of unusual financial market material cost fluctuation. Such added cost is called margin cost fluctuation addition.

It is also to be emphasized that the economic study shall take in consideration added factors of certain profits, which is limited by the reasonable life of the equipment used during the project life taking in consideration the residual life of the equipment (salvage values) and other factors, which are required during the maintenance of the operating of the project. It is well-known practice that the project total cost is governed by the owner's private staff to monitor the final cost estimate of the new project, and to submit such estimate to the board of directors of the project owner for final review and approval.

It is the board of directors' privilege to add what is necessary for the completion of the prospective project total completion of the feasibility study, and whence such final action is produced, and the board has approved the project on hand, then the study will be passed on to several outlet of engineering firms to bid on the execution of the detailed engineering work of the proposed project, and in some instances, the firm which prepared the feasibility study for the project on hand may receive the authorization of doing the actual detailed engineering work.

The aforementioned details are aimed to provide the young engineer with the practical steps that he needs when he/she is required to do such feasibility study; however, the young engineer must have absorbed the basis of such requirements during his academic years of study.

The feasibility study understanding is a recommended step for most of the representatives of all kinds of engineering firms and other firms to discuss with the graduating engineer when they come to several universities to interview prospective graduating engineers, and as such to discuss it further with young graduating engineers to determine the possibility of future offer of potential employment with such firms.

1. First Meeting with the Prospective Project Owner

The first meeting of the prospective project feasibility study with the client will discuss the aspect of the new project based on the scope of work of the new project, which was set up already by the project owner.

This step is taken following the selection of the engineering firm to do the feasibility study in addition to all the contractual steps including the cost of the study and the time it will take the firm to finish the study of the prospective project.

The project owner (client) will ask the representative of the selected engineering firm to read the scope of work in a very careful manner to make sure that if there is a need for any clarification or whatsoever, it should be addressed by the engineering firm by submitting a certain document to the client addressing the outstanding comments. Otherwise, the engineering firm will be considered to have understood the scope of work of the new project without any reservation.

2. Study of Project Requirement and Application

The assigned engineering firm starts the feasibility study within a time schedule as agreed upon with the project owner, and in accordance with the official list of the engineering branches which will carry the task of the project feasibility study.

The engineering firm later on distributes adequate copies of the project scope of work among those engineers who were assigned to prepare the entire feasibility study of the proposed project. Those engineers will review the project scope of work and the requirement of the proposed project. The individual engineering team of each specialty later on will start to carry its own task very effectively.

The architect and civil engineers of the engineering firm, after reviewing the scope of work of the proposed project, and after making a field visit to the new site of the project, will start to make a very close evaluation of whatever equipment will be installed at the new site of the project with the assistance of each engineering team assigned to the study.

The architect and civil engineers will start to prepare the elementary diagrams that show how the new equipment will be accommodated.

This step of preparing the elementary diagrams will be supervised by a project manager for the complete preparation of those elementary diagrams.

The project manager is usually selected based on his professional experience with the prospective type of the project at hands.

The project manager, soon after his review of the elementary diagrams is finished, will distribute the architect and civil engineering drawings among all other branches of the engineering team in order for each branch of the engineering team to start its own engineering feasibility study.

As it has been shown in this write-up, the intention is to give an overall picture of the requirement of preparing a feasibility study of a proposed project; therefore the presentation was a dense study only, which may help the young engineer to be familiar with a feasibility study requirement in the start of his professional life.

3. First Meeting of Project Study Team

The project manager of the engineering feasibility study announces the first meeting of his entire team and is usually attended by some of the proposed project executive members of the client company.

The aim of such meeting will be to review the general requirement of the proposed project and to get the impression of how the assigned engineering firm members will handle the study as it is desired to be.

The engineering firm members are selected to be very experienced members in each individual expertise, and as such, the entire team will be rewarded accordingly, especially when it comes to awarding the detailed engineering study for the real execution of the engineering study.

In general, an engineering firm is usually very keen for the team to be very effective, and as such the engineering firm may place all the different branches of engineering in one hall to give enough flexibility of cooperation of those engineers among each other.

The arrangement of having all branches of engineering in one large office has also a positive influence on the project owner that the engineering firm pays a great attention to have a close relation among the engineers of the engineering firm to produce a good engineering feasibility study of the prospective project.

The positive influence on the owner of the proposed project may lead the owner to assign the real detailed engineering of the project study to the same engineering firm who produced a good engineering feasibility study.

Upon the first completion of the feasibility engineering study, the engineering project manager of the engineering firm holds his first meeting with the engineering team in order to review the preliminary engineering drawings and to see that such a study did comply with the proposed project's scope of work that the owner had set up for the project.

The first meeting may result with some comments that need to be resolved before any meeting arrangement with the owner of the prospective project. Such comments may be needed to be discussed with the owner to be added to the study if the owner agreed to such comments.

It is emphasized at this stage that the engineering firm needs to be very positive in any remarks that may be produced during the feasibility study and be approved by the engineering firm project manager, which may require a close relation with the owner of the prospective project. Such a relation can sometimes produce several correspondences with the project owner, which can be documented accordingly in order save any additional cost of engineering works.

Upon the completion of the preliminary engineering drawings of the prospective project, the engineering firm will start to produce documented drawings for official use or approval by the prospective project owner.

Such completed drawings will be finally submitted to the project owner for final approval and used by the engineering firm group to proceed to complete the feasibility study.

4. Preparing Project Main Engineering Drawings

The main engineering drawings are considered to be the directives for every specialty of the engineering firm group. It will be the first step to enable every engineer in the group to proceed to the next step of defining that particular specialty engineering study requirement, be it materials location documents or any requirement to produce a final feasibility study by such engineering specialty.

Every engineer in the engineering firm group needs to be familiar with the elementary civil drawings and be familiar with the proposed project scope of work, a matter which will help that particular engineer to present a complete feasibility study related to his specialty of engineering.

As stated earlier, this write-up will concentrate on providing a detailed practical electrical engineering study; therefore, the electrical engineer's main task in the feasibility study is to collect all information related to the electrical load power supply requirement, and later to prepare the required information pertaining to such electrical load and the required drawings showing how the electrical load is supplied from different power centers. The electrical drawings will be called the main electrical single line diagrams, which cover the alternate current and the direct current single line diagrams.

Once that type of study is completed by the electrical engineer, the established drawings will be reviewed with the engineering project manager in order to make sure that the study has complied with the proposed project's scope of work, then the project manager approves such drawings ready for the project owner's final approval. Consequently, the electrical engineer moves to prepare the second requirement of the study.

It is very essential that the engineering firm must keep a close relation with the owner's engineering staff in order to produce a good study to meet the cost and time element requirements of the proposed project, which may be beneficial to both sides, and hence it may encourage the owner of the proposed project to assign the engineering firm later to prepare a complete detailed engineering study of the proposed project.

5. Prepare the Elementary Cost of the Proposed Project

Each engineering firm specialty team starts preparing the cost study in accordance with the requirement of the proposed project. Such study will start right after the complete review by the team engineering study by the project owner, and his approval of the study.

The cost estimate will be elementary study based on the engineering team's experience and close review of the cost estimate by the proposed project engineering manager.

Each engineering team makes sure of adding what is called a contingency cost to the elementary cost estimate, which ranges from 10%–20% of the elementary cost estimate, and such cost will have to be approved by the proposed project owner. The indicated percentage addition may cover the entire cost estimate of the proposed project, or it may cover certain areas of the proposed project, which will be subject to the owner's approval.

It should be interesting enough to say that the experience of the specialty engineer will be very effective and help the project owner to approve all cost estimate in order to make sure that the overall engineering feasibility study can be later a reality for further detailed engineering study.

The overall preliminary cost estimate is to cover the cost of the project site, project required equipment, the cost of erecting and constructing the total project, in addition to the cost of the construction management, which includes preparing as-built drawings and finally the cost of commissioning of the installed equipment at the final steps of the final status of the project.

6. Management Review of Feasibility Study with Engineering Team

The engineering firm is considered the main responsible team for the engineering feasibility detailed study, and as such this requires the engineering team to have full cooperation among each specialty team of the engineering firm on one side with each other engineering team of the proposed project owner on the other side, and such continued relation preferred to be through the engineering firm project manager. The results of such cooperation, and needed meetings between the two sides, have to be documented accordingly, especially if some changes in the feasibility study requirement took place but never deviated from the subject of the scope of work requirement.

If changes had happened in the feasibility study, which has deviated from the project scope of work, then such changes need to be reviewed by the proposed project owner, and the changes need to be addressed fully before implementing, and an official approval will take place in order to add the cost of the changes, to be compensated fully.

7. Engineering Feasibility Study Review with the Project Owner

The engineering firm is very much concerned to produce an engineering feasibility study to meet the requirement of the proposed project scope of work in addition to complying with whatever the project owner and the engineering firm have exchanged of valuable additions during the several meetings held between both sides to be included to be added to the original project scope of work.

The engineering firm also reviews with the owner of the proposed project all of the preliminary engineering drawings established during the feasibility study. It is to be emphasized here that the terminology of the preliminary drawings is used here because the study is a feasibility study, and as such the engineering drawings are called preliminary drawings.

The proposed project drawings become detailed drawings when the actual engineering study of the proposed project becomes a reality to evolve into detailed engineering study at a later date when the project becomes an actual executed project.

If the feasibility study review appeared to lack more requirements during the review, then the engineering firm will suggest to the owner some minor modification, and such modification may need additional engineering costs unless the engineering firm waives such costs, and this will encourage the owner to assign the detailed execution engineering to the same engineering firm.

It is assumed that the review of the proposed project may result in some engineering gaps, and such engineering gaps should be recorded in order to be resolved later, and another meeting with the engineering firm may be required and be requested by the proposed project owner.

The proposed meeting could be an overall review of the proposed project or can be just to review some of the changes, which took place during the last review of the project with the engineering firm.

The owner of the proposed project may require that the final review of the proposed project may be divided in accordance with the nature of the changes, which took place in a prior meeting time, and such request will be addressed at the start of the final review of the proposed project, and hence there will be a formal document addressing the owner's requirements.

It is needed to address the fact here that the engineering firm will be required at the final stage of the feasibility study to submit a complete engineering feasibility study. This final study should have taken care of the items that were discussed with the owner of the proposed project during the very last meeting. The action will avoid additional meetings, and the cost of any changes, or missed changes in the proposed project. The engineering firm should take in consideration that all changes in the study have been taken care of in order to avoid any misunderstanding with the owner of the proposed project.

The engineering firm project Manager will have to review with his engineering staff the feasibility study before any contact is made with the owner of the proposed project in order to make sure that the engineering feasibility study has taken care of all requirements of the project scope of work, and other engineering changes which have been approved by both sides to be included in the study. The manager then approves the study ready to be submitted to the owner of the proposed project.

The final engineering feasibility study will be submitted to the owner accompanied by an official letter of the presentation, and such a presentation may be done together with the presence of some lead engineers of the engineering firms.

The official letter will address the fact that the engineering feasibility study has been completed and has included the requirements of any changes addressed in prior meetings with the owner of the proposed project.

The official letter to the owner will request the owner to specify a time limit for the owner to accept officially the feasibility study, in addition to a final request of the engineering final cost of the study.

The final meeting between the owner of the proposed project and the representative of the engineering firm may include legal presence from both sides in order to declare the satisfaction of both sides, and such a meeting will be legally documented.

The final meeting between the owner of the proposed project and the representative of the engineering firm may be extended to include any possible agreement between both sides to assign the engineering firm to do the future engineering detailed study of the proposed project. The assignment will be based on the complete satisfaction of the prepared engineering feasibility study by the owner of the proposed project.

CHAPTER 2

Electric Power Generation Plants Review

I. GENERAL REVIEW OF POWER GENERATION PLANT

1. Conventional Steam-Turbine-Driven Power Generation Plants

The steam power has been used to drive certain mechanical devices called turbine, which is connected to a generator, to produce electric energy; and as such the equipment is called steam-turbine-driven power generator, which have been in use for a long time.

U.S. Energy Demand

America Is Projected to Need 50% More Electricity by 2025

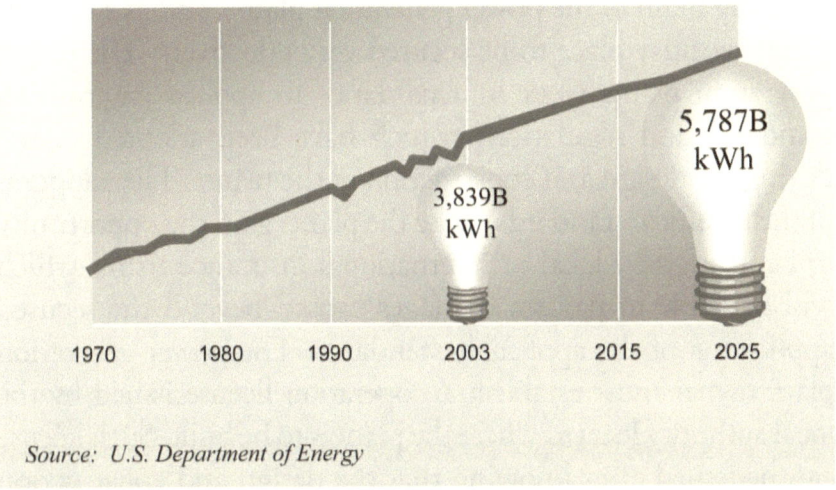

Source: *U.S. Department of Energy*

The steam power has been useful to move large generators to generate electric energy rated over 1,200 megawatts of electric power, and as such large steam turbine was used to generate such power. The conventional steam-turbine-driven power generation plants consist usually of several mechanical equipment, such as steam generated by the boiler, steam condenser, water pumps to pump water into the boiler, and cooling tower in addition to the stacker (chimney), and other auxiliary mechanical equipment. The power generation plants consists also of electrical equipment, such as the main power generator and its associated equipment in addition to the main power transformer and its associated equipment, and also the main auxiliary power transformer, and its associated equipment, in addition to auxiliary power distribution panels, and control panels. The other electrical equipment includes the direct current power supply (battery banks) equipment, and battery charger equipment in addition to the direct power control panels and associated converting power distribution power panels. The power generation plants consist also of the main control room and associated equipment, such as main control board and associated control switches and alarm system. The main control room is considered the directing brain of the power generation plants, and as such, it is a very sensitive place to be secured very effectively. The power generation plants work in accordance to applied engineering standards and regulations, which have been adopted to use during the design and construction of the plants. The adoption of the technical standards make the plants gain the opportunity to be insured by local or international insurance firms, which will be liable to pay for damages caused beyond the secured application of the applicable standards. The power generation plant owner must establish an operation license issued by the local authority located where the plant is to be built. Such license can be issued after knowing that the design and construction of the plant has adhered to all pertaining technical standards application, and such documents will be submitted to the local authority by the owner of the power generation plants.

The power plant insurance policy shall cover all human protection refund application, in addition to all the equipment, which have been housed inside and outside of the power generation plant location. Insurance firms are obligated to pay the owner of the plants all needed refunds should the plant be exploited in the insurance protection policy of the plants to enable the owner of the plant to put the plant back in operation.

Typical Model Steam-Turbine-Driven Power Generation Plant Equipment

1. Cooling Tower
2. Cooling Water Pump
3. Three Phase Transmission Line
4. Main Power Transformer (High Voltage Step Up)
5. Main Power Generator
6. Low Pressure Steam Turbine
7. Boiler Feed Water Pump
8. Surface Condenser
9. Intermediate Pressure Steam Turbine
10. Steam Control Valve
11. High Pressure Steam Turbine
12. De-Aerator
13. Feed Water Heater
14. Coal Conveyor
15. Coal Hoper
16. Coal Pulverizer
17. Boiler Steam Drum
18. Bottom Hopper
19. Super Heater
20. Forced Draft Fan
21. Preheater
22. Combustion Air Intake
23. Economizer
24. Air Preheater
25. Precipitator
26. Induced Draft Fan
27. Fuel Gas Stack

1-1. Power Plant Mechanical Equipment

The steam-turbine-driven power generation plants use the generated steam that the boiler produce as a result of burning petro liquid material or using burning gas material to heat the water in the associated water boiler.

Some other fuels, such as vulcanized carbon, was used in the past to heat the water boiler and to produce the required steam, which to drive the associated steam turbine as in the other fuel used.

The process of steam production in the power generation plants depend on the use of several certain mechanical equipment such as the burner, which uses several types of fuel, and in association of certain type of fans, (draft fans) to supply the burner with adequate burning air. The burner system is connected with the main boiler equipment, which produces the required steam to drive the main turbine and, in turn, to drive the main generator to produce electrical energy.

The main boiler is also connected to a system of boiler water supply pump to supply the boiler with water, and when the water is adequate then it will be boiled inside the boiler to produce the driving steam to the turbine.

The burnt fuel (smoke) out of the boiler is driven later to an element called the stacker chimney. The fuel used in the boiler system can be petroleum, gas, or vulcanized carbon. The main turbine that is connected to the boiler is equipped with several blades to receive the produced steam with certain driving power pressure to drive the turbine; however, prior to reaching the blade of the turbine the steam runs through several sets of controlling valves. These valves are the block valves, shutdown valves, and control valves. The valves' job is to control the amount of steam entering the turbine and other valves to stop the turbine. The control of the turbine speed is governed by a device called the governor). This device sends control signals to the previously mentioned valves to open or close those valves as required to maintain a certain speed of the turbine.

The blades of the main turbine are usually mounted on an axle of the main turbine called turbine main axle, which rests usually on certain bearings enabling the entire turbine system to turn, and the system is kept turning by associated electric motor, which helps the turbine axle to stay turning even at rest to avoid any bending to the main axle because of the heavy weight of the turbine, in addition to a lubricating pump to lubricate the action of the bearings.

The design requirements of the size of the turbine depend mainly on the size of the power generation plant. There are small plants that require one turbine set, and other large plants that require a chain set of turbines. Such turbines are called high-pressure turbine, intermediate turbine, and finally, low pressure turbine. Those turbines each provide the required action to drive the main generator of the power generation plant. The turbines are connected to the associated condenser to receive the condensed unused steam, which later this steam will be driven to filtering equipment, and to return as liquid way back to the original boiler after it passes a thorough special heating elements called heater, which gain its heating action by the gases emitted by the boiler burning fuel as shown in the typical operational.

Rankin Heating Cycle
1 to 2: Isentropic expansion (Steam Turbine)
2 to 3: Isobaric heat rejection (Condenser)
3 to 4: Isentropic compression (Pump)
4 to 1: Isobaric heat supply (Boiler)

It should be understood that the operation of producing steam from the boiler is for the steam to drive the turbine that was described earlier. This operation is more involved operation, and the purpose of summarizing it here is to give the young electrical engineer more general experience to aid him/ her in preparing the required electrical study for the mechanical equipment which is part of the power generation plants.

The steam turbine is connected mechanically to the main generator, which is the main electrical equipment that all the mechanical equipment work for in the power generation plant.

This type of generator is called synchronous generator, and it is usually a three-phase power generator that is connected from one side to a system that is called isolated phase bus bars for cooling purpose of the generator bus bars of the power generator.

The bars are located in special containers called bus ducts, which is isolated in a special way to permit a cooling effect to cool the electrical bars during operation. The entire cooling system of the main electric bars is called bus duct cooling system, which needs required power supply to run the cooling system.

The isolated phase bus bars are connected to power cables through power circuit breakers called generator main circuit breakers. These circuit breakers are usually located in a distribution area called main switch yard, and sometimes those breakers are located in the main steam drive turbine area in accordance with the desired equipment and the engineering study to deliver the generated power. The desired engineering study also makes available certain power breakers and control systems together with power cables to transfer such power outside of the main turbine-generator area in accordance with applicable standards.

It is stated again that all of the preceding information is general information in order to help the design engineer with his electrical or mechanical study; however, there are other auxiliary equipment the engineer needs to be familiar with, and I will refer to such equipment during the text of this write-up.

It is possible that certain power generation plants may have the main equipment (turbine + generator) inside of one main hall called (turbine-generator building), and it is possible that such equipment may be located in an open hall.

1-2. Power Plant Electric Equipment

1-2-1. Engineers considered that the main power generator is the main electric equipment in the power generation plant despite the fact that the mechanical equipment play a major role in the power generation plant.

The main power generator is usually set next to the steam-driven turbine and are mounted on the same table, which is called the turbine-generator pedestal.

The pedestal is usually built out of a combined armed concrete table at no more than two meters high, and it is built in accordance with certain engineering standards to carry the weight of both the turbine and main generator weight, in addition to that of seismic requirement to withstand the shake of the equipment mounted on top that table during the normal operation of the turbine-generator.

The main power generator is a special equipment called synchronous generator, which consists of two items working together; one item is fixed to the body of the generator, and the other item is rotating because the action of the associated the steam turbine driven to produce the generated current. One item is called field structure, which produce the magnetic field of the main generator, and the other item is called armature, which consists of a group of special wound electric wire carrying electric current.

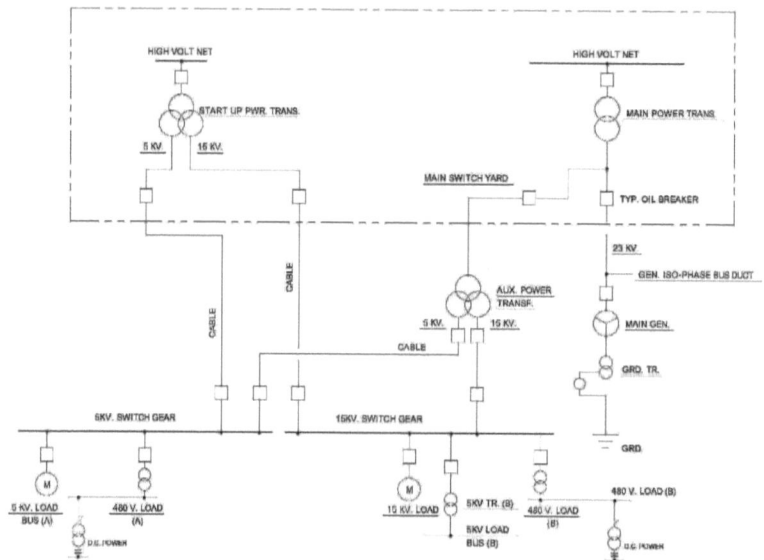

Conventional steam-turbine-driven power generation plant Electric single line diagram (diagram showing power distribution scheme for the plant).

A generator usually is a rotating element that receives the rotating action from the moving turbine, while the other part of the generator is fixed to the body of the generator to deliver the generated current to the electrical load which the generator supports.

1-2-2. Electric Power Transformers, Circuit Breakers, and Generator Power Bars

The main power generator is usually connected to the electrical power transmission lines through the generator main power circuit breaker; however, prior to this connection there will be two electric feeders connected to the main power generator, namely, the first electric feeder supplies power to the power transmission lines through associated oil power circuit breaker, and the second electric feeder supplies power to what is called an auxiliary equipment power transformer, which is connected to the auxiliary equipment power supply panels.

All of the mentioned power circuit breakers are the oil type circuit breaker and are located in the main switch yard of the power generation plant, which is located near the power generation plant. A typical power transformer is shown below:

Typical main power transformer.

It is to be mentioned here that while the main power generator is in the process of preoperation (not connected to any electrical load) in the startup mode, there will be another similar main power transformer called start-up power transformer. This transformer is connected to the power transmission lines to provide a source of power in order to supply the plant with power source to run the plant auxiliary load during the startup mode of the plant. This startup transformer could exist at a distant location near an operating power network, or from a new power transformer located at the new power plant location, to be powered from the power transmission lines which may be owned by the same owner of the new power generation plant. This means that the new power generation plant could have three power transformers at its new location.

These transformers, and the new main power generator, are supplied with a grounding system in accordance with applicable standards stated by the local authority where the new power plant is built.

The grounding system of each electrical equipment in general may vary as stated in the applicable standards. The power transformers may be grounded directly (grounded neutral) while the main power generator is grounded through associated equipment called generator neutral grounding, which is a part of the protection provided to the main power generator.

The auxiliary equipment power requirement is derived from power supply panels, which the Auxiliary power transformer supply with as soon as the feeder from the main power generator is available, and through a power transfer scheme to transfer power source from the startup transformer to the auxiliary power transformer. The main power generator required protection will be discussed in a later section, which will discuss all required electrical equipment protection.

The main power generator usually works with the aid of the steam-driven turbine, which is the mechanical mover element, and also with the help of control panels called generator excitation control panels. The panels also control the requirement of the magnetic excitation, the control panel of cooling the generator field, and the control of the generator main electric bars, which is called the generator field and isolated phase bus bars cooling system. Control Panels are usually located near the main power generator.

As stated earlier, the main power generator supplies power to all equipment which require power supply, and the required power is usually supplied through power panels called auxiliary equipment switch gears.

The generator is also connected through associated oil circuit breakers located outside of the plant area called the main switch yard.

Some of those oil circuit breakers are used partially to power plant auxiliary loads, and other oil circuit breakers are used to connect the generator to power transmission lines to supply power to consumers via the main power transmission lines.

It is imperative at this point before going through the details of the variety of required type of power distribution panels (power switch gears) to describe the means of power supply to those panels.

The generated power is usually sent through oil power circuit breakers located in the main switch yard, and then part of the power supply is sent through associated circuit breakers, and the main power transformer for use by the consumer, and the other part is sent also through associated circuit breakers, and a power transformer called auxiliary power transformer. See typical picture above.

The main power transformer consists of two windings called the primary windings and the secondary windings. One winding is connected to the main power generator, and the other winding is connected to the power transmission lines.

The auxiliary power transformer consists of three windings (the primary windings), which is connected to the main power generator and two secondary windings, which are connected to two different power switch gears, and the figure also shows the main switch yard where generated power is connected to associated power transmission lines and other electric connection.

Power and control cables are usually sized in accordance to IPCE standards using cables schedules.

It is to be referenced here that there are two types of electric connecter bars associated with the main power generator.

The first type of the bars are used to connect the main power generator with associate loads through what is called generator isolate phase bars while the other type is located outside of the power generation plant in the main switch yard for distributing power from the main switch yard to specific consumers electric loads.

The isolated bars are supplied with a special cooling system, which is called the isolated phase bus duct cooling system.

Electrical engineering studies are divided between two teams of engineering firms. One firm prepares the electrical engineering study for the inside of the power generation plants, while the other firm prepares the electrical engineering study for the outside of the plants, which is called the power distribution and transmission lines study. It is very possible that one engineering firm can do and be assigned both studies work according to the qualification of the engineering firm.

The generated power out the plant is usually connected to certain bars at the main switch yard and then carried from the main switch yard by bare wires through oil power circuit breakers. All power distribution emitted from the power generation plants must go through associated oil circuit breakers for protection and isolation purposes.

1-2-3. Power Supply Distribution Panels

Power generation plants are usually equipped with mechanical equipment tied to electric motors. These mechanical equipment are called auxiliary equipment.

The mechanical equipment associated motors which connected to associated power supply distribution panels are called switch gears or power switch boards). There are also other auxiliary control panels that require electric power. The required power for both types of services are usually connected to the auxiliary power transformer through associated power circuit breakers during normal operation, and they will be connected to the startup power transformer during the startup of the power generation plant. The aforementioned switch gears and power switch boards each consist of electric horizontal bars as well as electric vertical bars, which are interconnected with each other to supply the associated circuit breakers contained and isolated within the associated enclosures of the panels.

The circuit breakers in the switch gears or in the power supply panels are connected to the moving motors load of the auxiliary mechanical equipment (usually pumps or fan motors) connected through specified isolated electric wires. All circuit breakers that are connected to moving electric motors are supplied with associated control switches placed on the surface of the switch gears or power supply panels (Sometimes called switch boards), and sometimes additional remote control switches will be required and placed on the main control board in the main control room, or at other local control panels.

It is to be stated here that the specialized engineer (instrument and control engineer—a member of the engineering study firm) will designate the location of all required switching devices.

The power supply panels (switch gear or switch board) will include all power circuit breakers to operate and protect connected motor loads and other feeders in addition to include also all required protection devices such as protective relays, and operation indication lights, and indicating reading instruments to be placed on the front surface of the power supply panels.

The protective relays requirement, and the instrumentation and process control equipment will be discussed later in this write-up.

Power supply panels' varieties are normally located in the close vicinity of the electrical load that they supply power with.

Those locations can be in the turbine-generator main hall or at specified convenient locations near the motors load, while the location of the main power transformers should be at the main switch yard. The location of the control of associated breakers of the main transformers will be placed on the surface of the main control board in the main control room, in addition to locations on some local control panels near the main switch yard.

It is to be emphasized again that all power transformers' primary and secondary windings are protected and controlled by oil power circuits located at the Main switch yards.

The Switch Gears and control are protected by air power circuit breakers located inside the switch gear panels in addition to control switches, and circuit breakers position indicating lights, which will be mounted on the surface of the switch gears. All position indication lights of the aforementioned circuit breaks will be duplicated and be mounted on the surface of the main control board. The indicating lights, which are mounted on top of the associated control switch of the particular breaker, such indicating lights lenses are colored, one with red color to indicate the closed position of the breaker, and second light colored with green color to indicate the breaker is in the open position, and the third light color is amber color to indicate the breaker in the outdrawn out of its cell position. The coloring of the indicated cells shall be in accordance with the local applicable standards, and that may vary.

1-2-4. Essential Auxiliary Power Supply Sources

The power generation plant has also an auxiliary power source that powers certain equipment called essential equipment, such as the turbine axle turning gear motor and the turbine axle bearings oil pump motor and other essential equipment. The power source usually consists of the direct current power supply panels connected to battery banks, which is connected to the battery-banks charger system connected to an alternate current (AC) power panel for current conversion.

The battery-banks supply direct current to panels consisting of the main circuit breaker and branch circuit breakers to carry the power to the aforementioned motor loads.

There is also an additional system called essential power supply or uninterrupted power system (UPS), which is used to power the instruments control requirement and other requirements.

The UPS system consists of an AC source, one system of detecting the interruption of the AC source, battery bank, and one inverter to change the battery-bank direct current to AC current, in addition to a distribution panel to distribute the uninterrupted AC current to power the essential instruments control requirement.

1-2-5. Power and Control Cable Carriers

There are two types of cable carriers used in the power generation plants. One type is called cables trays, and the other type is called cable conduits. Most of the cable tray carriers are made of metal, which can be the galvanized metal or aluminum metal type. The carriers are used to carry cables from the power supply centers or from control panels to the associated electrical loads. Cable conduit carriers are similar to cable trays (cable carriers), are metal or a plastic type (PVC conduits), and are, to a certain extent, similar to pipes, which carry water or steam.

The design of routing any cable carrier is prepared by a certain study, which is associated with the requirement of the powering and control of electrical equipment in the power generation plants.

The power generation plant requires also certain cables for the use and control of instruments (instruments control cables), and as such certain cable carriers will be assigned for this type of services. The assigned carriers are prepared to transport instrument cables from the associated control panels to the instruments location throughout the plants as dictated by the design requirement.

Cable trays are usually run over ground horizontally or vertically, while cable conduit may be run over or underground. Conduits, as described earlier, can be metallic or plastic (PVC conduits). When metallic conduits are used, and buried underground, the conduits should be buried under a layer of forced concrete layer for protection. It should be noted here all described carriers can be in different sizes in width (trays) or in diameter (conduits) to accommodate a limited volume of cables as required by applicable standards for routing cables.

Cable carriers' paths can be routed through the use of certain computer programs or the use of the manual conventional way of routing.

Cable trays physical arrangement is made in accordance with the volt potential capacity of the cables, which the cable tray will be carrying.

Cable trays usually are lined and carried on top of each other with a defined separating space to ease installing carried cables. The lineup of cable trays have to be stacked on top of each other, arranged so the highest cable voltage level of the carried cables tray will be in the most upper tiers of the trays line up followed downward with the second cable voltage level tray, and end up down further with the lowest voltage cables tray, which is called the instrument and control cable tray).

All cables, when routed out of the power supply panels, such as power cables or control wires, are usually sent through associated conduits below till the cables reach associated cable trays. The cables will then be routed through associated cable trays till they reach the final destination of cables, which could be motor loads or controlled instruments, then the cable are run again through associated conduits to finally be tied to the terminals of the associated motor or tied to associated instrument. The running cables volume inside associated conduits or associated cables trays should comply with the applicable local standards, and in general should not exceed the volume of the its carriers.

The detailed engineering study of cable routing and cables volume determined values will be discussed further later in this write-up. It is further stated that cables selection, whether they are instrument control cables, power control cables, or power cables shall be selected in accordance with applicable standards, such as standard IPCE (US standard) or applicable local standards, and this will be related to cable cross-section area, cable insulation materials, and required cable protection.

1-2-6. Battery-Banks (Direct Current) Power Source

Battery-banks power source play major roles in the power generation plants. This type of power source is used to supply essential load power supply as indicated earlier. This power source is applied using two different potentials, namely the 220V (direct current potential) to power direct current motors load, and 120V (direct current potential) to operate certain type of air circuit breakers), which use this type of potential to operate such circuit breakers in addition to a lower direct volt of 20V (direct current potential), which is used to operate instruments power supply. Furthermore, the direct current power supply is used for operating the UPS System where direct power source is converted to alternating current (AC) source for powering certain instruments with uninterrupted power supply.

Usually battery-banks power source with its charging equipment, and associated direct current power source distribution panel are all installed in a special room, which is completely aerated in addition to associated maintenance equipment and protective clothing and a water source to wash out any harm caused by operating the direct power source equipment to be located in an isolated room from the battery-banks equipment for safety precaution purposes. All direct power source equipment selection and safe protection must be in accordance to local applicable standards and fire protection code and regulation requirements. In general such equipment shall be selected to be explosion-proof equipment type per required local standards.

1-2-7. Plant Instrument and Control Panels

Power generation plants require certain type of instrumentation devices (control instrument), which carry electrical signals (control orders) between certain electrical loads and instruments mounted on the local control panels or instruments mounted on the main control board in the main control room in order to perform in accordance with the requirement of the power generation plant (mechanical process) operation. The plants also are controlled and protected by electrical instruments, such as control switches, protective relays, and indicating lights, which carry electrical orders to turn off any electrical loads, and to disconnect the associated mechanical loads from the power source. These control instruments are usually mounted on the surface of its associated power source panels (switchgears or power panels) and are duplicated on the surface of the main control board in the main control room.

1-2-8. Communication (Hearing or Indication) Instrumentation

The hearing or indication instrumentation are very important devices in the power generation plants operation, as they are the critical communication among all operating men medias for the safe operation of the power generation plants. The type of instrumentation is usually hearing devices, such as telephone devices or public address system, which operate throughout the plants and is communicated back to the main control room.

However, the indication instrumentation is usually the monitors type, television type, or the annunciator type to give a flashing signal to alarm the operator in the main control room, or local control panels to take a protective action of operation.

1-2-9. Fire Protection System

The international and the regulatory standards and Fire Protection Authority require that the power generation plants must install a fire protection control system in every operating plants for the sake of protecting working personnel and equipment in order for the power generation plants first can purchase insurance policy to protect all working personnel and protecting the plant from a partial loss or complete loss of equipment in the power generation plants. Fire protection can be through the use of gas type to stop fire by eliminating burning fire and isolate it from further oxygen supporting fire or using foam to prevent fire extending to other close by areas in the plants.

1-2-10. Power Generation Emergency Exit Signs

It is an emergency regulation standard for power generation plants, and very forced regulation, to install emergency-exit-lighted signs in every industrial plants whether it is a power generation plants or any industrial plants in order such plants to have permit to operate its facility prior to any startup of normal operation. Furthermore, such emergency exit lights location in any plant is defined by the local authority where the plant will be operating. The signs are designated to help a safe exit people who work at such plants including public people who are allowed to visit the plant.

1-2-11. Power Generation Main Control Room

Main control room is considered to be the managing brain of operating any power generation plant where essential control instruments are located on both the vertical portion, and the horizontal portion of the main control board, in addition to major control switches, and indicating instruments, and the main mimic bus, which indicate the plant mechanical process sequence of the plant operation, add to that the location of the plant (annunciator system), which alerts the operators in the main control room about any malfunction of major plant items that may jeopardize the plant normal operation.

2 Conventional Gas-Turbine-Driven Power Generation Plants

The use of gas-turbine-driven power generation plants was used and spread out after the discovery and use of the liquid natural gas, which was pumped out first before refined together with the petroleum liquid. The released gas out of oil wells was called sour gas. However, the mixture of the sour gas with the liquid oil was separated apart, and the sour gas was released to be burnt out in the atmosphere. People used to see the big flame of the burnt sour gas while the unrefined liquid oil was pumped to be exported throughout the world, except a small portion of it was refined locally in some of the Arab world for local use as heating oil and other uses. Soon later on the refined sour gas, called LNG (liquid natural gas), was exported out through the Arab World, and other countries to have become the second fortune for most of the Arab World.

The discovery and use of the (LNG) gas encouraged other discovery, such as the natural gas mines, and the volume of such substance became a third fortune of many countries, including the Arab World. The discovery of the natural gas had encouraged many international media to use such gas as driving force in some types of turbines called gas turbine, and the turbine was coupled to electric generator to produce electric energy in areas where water quantity was limited. The new facility (turbine-gas-driven power generation plants) had helped to use such facilities as mobile facility which can be moved to where it is needed the most.

The gas-turbine-driven power generation plant consists of the gas-turbine-driven equipment, which tied up to the driven main power generator, in addition to other mechanical equipment all located in a limited space to make such unit to be gathered with other gas turbine units in a total area. All those units are usually controlled from one location called the main control room, which has a main control board to control several unites of the described facility. The electrical equipment generally consists of the main power generator and power distribution panels in addition to the power distribution yard, which is called the main switch yard. Most of the gas-turbine-driven power generation units have a power generation capacity of 5 megawatts to 45 megawatts.

The future trend is to extend the present power-generating capacities into higher capacities within the limitation of the safety of human operators and the safety of the gas-turbine-driven power generation equipment, which are matters of great attention for a better future and application.

The gas-turbine-driven uses the theory of pressurizing the air in a vessel then the compressed air is heated in a combustor chamber, after which the pressurized heated air will be guided through certain passes to reach the turbine then the air will be inflated inside the turbine body to push the turbine fins into a continuous circular motion to turn the attached generator into a motion of generating power energy as shown in the symbolic figure below. The entire cycle of the motion is called ideal Brayton cycle.

It is worth to mention that the exhausted unused heat, which could be let out directly to open air, can be harnessed before it was lost, can be used again to warm out the inlet air to the compressor, and as such this will improve the efficiency of the gas turbine operation. It is further known that the exhausted gas, after it has been used in turning the gas turbine, goes through a certain equipment called waste heat recovery unit to improve the gas turbine's overall operation efficiency as explained earlier, or it could be used to heat the spent steam in the operation of the steam turbine, which also improves the efficiency of the steam-driven turbine in such power generation plant.

The overall efficiency of the gas-driven turbine operation may not exceed the 33%; therefore, the excessive exhausted heated gas, which may reach the 67% of the total hot gas, is actually considered a heating loss unless it is used, as was explained earlier, for heating the exhausted steam in the steam-driven turbine operation. The following figure shows the heat percentage loss and the percentage of useful heat cycle.

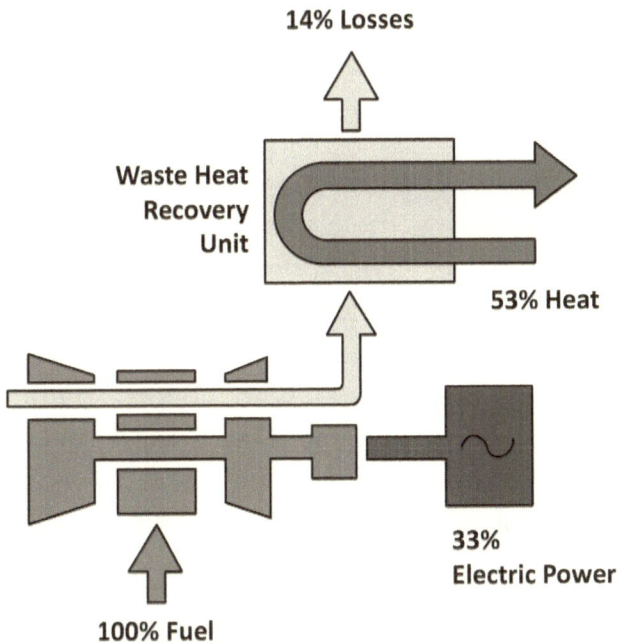

Flow chart illustrating the cogeneration principle

2-1. Gas-Turbine-Driven Mechanical Equipment

The gas-turbine-driven mechanical equipment consist of the turbine body and other mechanical equipment, such as the turbine moving turning gear and associated electric motor, in addition to the gas fuel pumps, which are in accordance with the turbine type as shown in the figure below, added to that the heated fuel (combustor) and the waste heat recovery unit, plus fuel tank, and last is the chimney.

The gas turbine equipment is mechanically considered less complicated than the internal combustion engine equipment because of the simple structure of the total gas turbine unit because of the assembly turns all in one unit (axle, combustor, turbine, and associate generator) stand as one unit except for the fuel tank. However, the entire unit gets more complicated when the turbine gets larger, and as such the turbine speed gets to be slower, and certainly the axle bearings get more complicated in accordance with the size of the gas turbine and associated generator.

2-2. Gas-Turbine-Driven Electrical Equipment

Generally speaking the electrical equipment in the gas-turbine-driven power generation plant is smaller than the electrical equipment used in the steam-turbine-driven power generation plants namely because the main power generator size in the latter is bigger, and it is an independent unit, which requires special tying equipment with the associated steam turbine during the power plant construction, while the gas-turbine-driven plant the entire unit comes as an integral unit:

Gas-Turbine-Driven Generator Assembly

However, the power generator structure in both types of the plants stay the same (synchronous generators) except for the difference in the kilowatts capacity, which is smaller in the gas-turbine-driven power generation plants. Furthermore, the electrical equipment are very similar (electromagnetic excitation requirement) to each other, and so is the requirement of the generator (neutral grounding), and other requirement that are listed for the power generator of the steam-turbine-driven plants.

2-3. Plant Power Transformer

The gas-turbine-driven power generation plants usually use one main power transformer (see typical figure) in order to transmit electric energy to consumers. The potential levels of the power transformer usually ranges from 66 kilovolts at the power generator terminal to step up till 210 kilovolts at power transmission lines level in accordance with the design requirement of the power plants.

As stated earlier the gas-turbine-driven power generation plants are used in places where water availability is limited for the use of steam-turbine-driven power generation plants. Sometimes the gas-turbine-driven power generation plants are used where only limited power is required for operating privately owned facilities, and as such there is no need to use power transformer. In that case, the generated electric energy is directly connected to power distribution panels at the specified facility.

Small gas-turbine-driven-power-generating units have voltage terminal rating ranging from 11 kilovolts to 14 kilovolts, and the range may differ in accordance with some manufacturing companies' designs; therefore, there will be a need to use certain small power transformers to lower the generator terminal voltage to a suitable distribution level as required by the electrical load at the consumer plant.

2-4. Power Supply Distribution Panels

The power distribution panels used in the gas-turbine-driven power plants are similar to the ones used in the steam-turbine-driven power plants with regard to powering requirement to power associated auxiliary load. However, the operating potential my differ in accordance with the size of the auxiliary load, and as such there will be a need in the steam power plants to use auxiliary power transformer to provide the right potential at the auxiliary power transformer secondary windings to suit the connected electric load.

2-5. Auxiliary Power Sources

The auxiliary power sources is a requirement in every power generation pants, and as such there will be direct current (DC) power source (battery-banks), and associated equipment, there will be an essential power source (DC source) converted into (AC source) in the operation of the gas-turbine-driven power plant to aid in the operation of the plant also as it is in the steam turbine power plan.

2-6. Battery-Banks (Direct Current) Power Source

The battery-banks power source in the gas-turbine-driven plant play a major role in the operation of the plant, and the battery-banks vary in the potential it is used, which is 12 DC volt to 125 volt. The battery-banks are charged using a special charger, which is connected to an AC power source outlet supplied by the plant generator. All of the battery system and associated equipment shall be installed in a private room, and all battery equipment shall be explosion proof to minimize the effect of the gas emitted by the batter-banks equipment.

2-7. Instrumentation and Control Panels

Instruments and control panels in the gas-turbine-driven plant is less complicated than the steam-turbine-driven plants. It is usually several such plants control instruments are gathered in one location at the main control room of the several plants.

2-8. Electric Cable Carriers

Cable carriers in the gas-turbine-driven plants are not much different from the carriers in the steam plants, and they are less complicated; therefore, there is no need for a computer program to do the trays' automatic routing, nor is it required for computer (automatic cable routing). However, the selection of cable carriers, such as cable trays and cable conduits are done the same as in other like in other power-generating plants, and cable carriers are separated so the high voltage cable tray location will be in the most upper tray position followed with trays with lesser voltage level and low-voltage instruments will be in the lowest position of the trays tier.

2-9. Plant (Communication and Indication) Instrumentation

Every industrial facility must have hearing and indication instrument provisions. The telephone system, together with loud speakers, public address, and television systems are essential equipment for the operation of the average and large turbine-driven-power generation plants. These types of equipment were fully described in the steam-driven power generation plants, and the description of such requirement may be described later in this write-up.

2-10. Fire Protection Requirements

Fire protection requirements are very important in every power generation plants and are enforced by the local authority where plants are built and operated, and by the insurance agencies, which provide man and equipment damage protection. Generation plants that do not have insurance protection will not be permitted to operate.

2-11. Power Generation Plants Emergency Exit Signs

The matter of providing emergency exit signs is as important as any required service in any power generation plants; however, this type of equipment is less complicated in its operation. The operation of this system is divided into two separate locations of the plant; one location will be to cover the location of the turbine-generator unit's area, and the second location will be the area of the main control room area where all of the turbine-generator units will be controlled from. Local authority at where the generation plants are built may also impose certain private rules of people evacuation during emergency situations in addition to the requirements of installing exit signs, and the specified requirement application will be reinforced in order to have the required insurance protection.

2-12. Generation Plants Main Control Room

Generation plants main control room contains all required control of the entire installed gas-turbine-driven generators assembly, which are built in one location. The main control room is used usually to control several gas-turbine-driven generation units in addition to local control for each generating units.

The main control room my also include provisions to control tie the individual generating unit power network as is dictated by the level power demand of the network.

The main control room has an alarm system (annunciator alarm system) to monitor any emergency condition situation that the generating equipment may have, in addition to what is called a mimic bus monitoring the power flow condition of all generating units tied to the power network. The main control board at the main control room may have all required switches to operate certain oil circuit breakers, which are usually installed at the main switch yard for the safe operation of the power distribution, and supply to the consumer with adequate power use.

3. Conventional Hydro-Turbine-Driven Power Generation Plant

Hydro-turbine-driven power generation plants receive their driving power from the moving energy of water, which comes from raindrops collected and accumulated behind tall dams. The natural collected energy as such, which created water lakes in some location, encouraged some utility companies to build water dams to harass the water lakes, and to take advantage of the collected water for the purpose of a possible future location to build on a hydro-turbine-driven power generation plant nearby the dam area.

However, it was found that part of the water source may be lost due to evaporation process, but the large part of the reserved water, the so-called water lake, stays adequate for use and to build a new hydro-turbine-driven power generation plant.

The hydro water source in general goes through several nature activities, such as volcanic activity and air current cycle activities, which the latter drive the ground water during cycles called hydrologic cycle to result into water drops to return to earth as falling rains. These hydrological cycles are very important in the operation of the conventional power generation plants; call it steam-operated or hydro-operated plants, which use water in their normal operation. Figure below shows and illustrates the sequence of the described hydrologic cycle.

1. The sun heats the ocean.

2. Ocean water evaporates and rises into the air.

3. The water vapor cools and condenses to become droplets, which form clouds.

4. If enough water condenses, the drops become heavy enough to fall to the ground as rain and snow.

5. Some rain collects in groundwells. The rest flows through rivers back into the ocean.

Hydrologic cycle

The following demonstrating figures show below figuratively the relation among the equipment involved in the operation of the hydro-turbine-driven power generation plant. The figures shown (water reservoir, dam, water intake structure, and control gate, and finally the canal Penstock) leads the water to drive the main turbine that is tied to the main power generator.

Hydro-turbine-driven power plant figurative operating relation

Figurative equipment inside hydropower plant

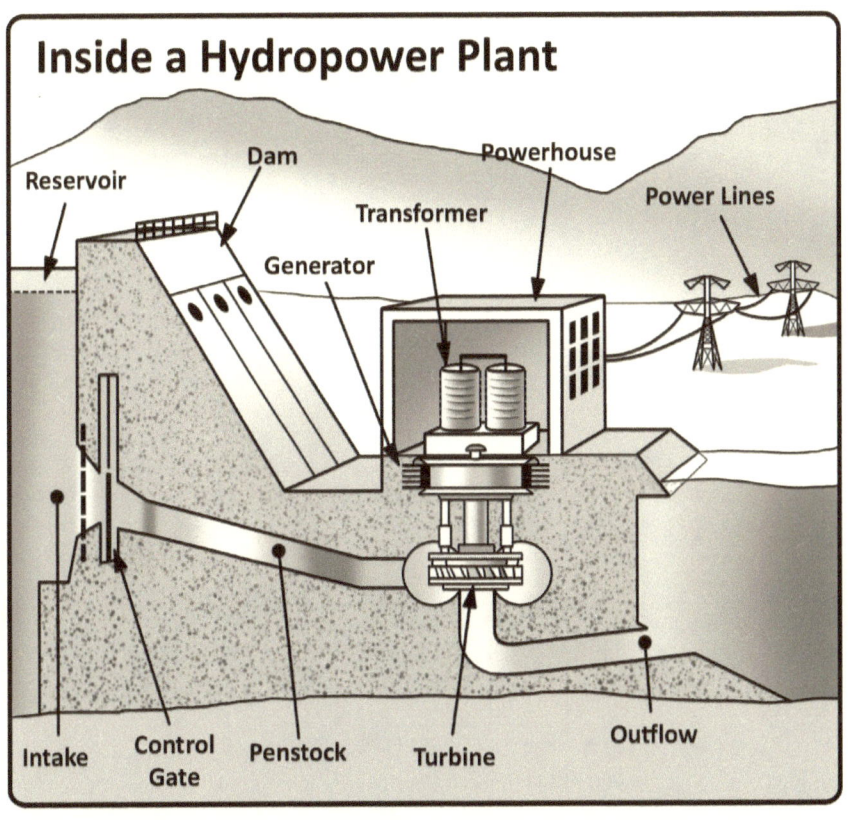

Figurative typical hydro power generation plant

Turbine-generator connecting shaft

Hydro-turbine-driven power generation plant depends on the moving energy of the water accumulated behind associated dam where water is directed to flow and reaches the turbine to produce a moving energy to turn the turbine, which is connected to the main power generator, and hence producing electric energy.

Hydro-turbine-driven power plant figurative operating relation

The flowing water through the Penstock whence it reaches the inner concaved blades of its associated turbine makes the turbine start to create a circular motion of the turbine plates structure, which is tied mechanically to the main generator shaft, and the entire assembly of the turbine–generator is set in motion for the generator to start producing electrical energy. The energy will be transmitted through power transmission lines to be delivered to the consumer.

The used water to turn the turbine are circulated through associated pipes called tail race ways, which carry the used water to a river directly located behind the lake that supply the useful water to the hydro turbine. It should be stated here that the heart of this type of power plant is definitely the main power generator. The following figures listed below show an assembly of several turbine-generator units, which combined to produce a desired level of electric energy at some specified power generation plant. As stated earlier, this type of power generation plant is located in one location, which is similar to the Hoover Dam location in the USA, where there are seventeen turbine-generator units are driven by water force accumulated behind the dam; however, there is a much bigger plant (Itapúa Plant), which is near the border of Paraguayan-Brazilian, and is considered the largest hydro-turbine-driven generation plant in the world, which generate electrical energy close to 75TWH (75 trillion watt hour) per year.

The elevation of the aforementioned dam reaches 196 meters high, and the length of the dam is 76.7 kilometers long. The length of the lake, which extends behind the dam, is almost 170 kilometers long, and it contains close to 29 billion tons of water. The number of power generation units at the specified location may reach 18 units. Each power-generating unit has a capacity of 715 megawatts. Each hydro-turbine unit receives 700 cubic meters per second of water-driving energy with an efficiency of 8.93%.

Most hydro-turbine-driven power generator operates as described earlier. However, there are some hydro power generation plant types that are pumped-storage types, which have two water storage facilities located near the hydro power generation plant, one water storage area called upper reservoir located behind the plant dam, and the second water storage area called lower reservoir, which receive the water coming out of the plant instead of wasting the water to pour in the neighboring river.

The general trend to use hydro power plants is when there is a big demand for electric energy. When high electric energy demand is not required, some or all the operating turbines at the plant will be operated in a reverse direction (reversible turbine) to carry water from the plant (lower reservoir) to return it back from the lower reservoir back to the upper reservoir to operate some of the hydro turbine units in order to move some of the power generator to produce electric energy when it is needed.

3-1. Hydro Power Plant Mechanical Equipment

The hydro turbine unit in hydro power generation plant is considered to be the largest equipment in the plant. The turbine assembly is different from other turbines used in other power generation plant because the blades in the hydro turbine are mostly more concaved than in the other turbines.

This way the hydro turbine can scoop a larger quantity of working water to move the plant associated to the big turbine. The figure below shows the picture of some of the generators used in the hydro power plant.

It was stated earlier that hydro turbine can be used in a reversal motion operation; therefore, the reversible operation needs special gears and moving electric power in its mechanical move. The electric power is usually supplied by the plant to drive the electric motor. The assembly (gears and motors) and lubricating oil pumps are considered parts of the mechanical equipment used in the hydro power plant.

The hydro power plant uses also a variety of control valves to direct the flow of operating water as required to ensure a safe generation of electric energy throughout the hydro power plant operation.

As mentioned earlier the hydro turbine is the largest mechanical unit in the plant and may weighs 172 metric tons); therefore, it requires a controller called the mechanical governor to control its speed during the normal operation of the hydro turbine. This is done through the use of hydraulic system to open/close certain control valves upon receiving control signals from the mechanical governor, and such action consequently regulates the generated electric energy flow of the hydro power generation plant.

The hydro turbine is usually located under the power generator unit, and the two units are connected through a big mechanical shaft). The big turbine and its associated shaft are located in the lower flat of the hydro turbine building while the power generator is located in the upper flat of the building, and this is done because of the safety precaution to isolate the wet mechanical equipment from the electric operating equipment.

3-2. Hydro Plant Electric Equipment

The main power generator in the hydro power plant is considered the heart part of the plant and used as one of several generators in the assembly of several power-generating units to produce electric energy collectively with other units in the same plant.

It has been described earlier that the Hoover Dam hydro power plant is considered the largest hydro plant in the USA, which generate power at 16,500 volts potential level. The potential level is sent through the main power transformer, which raises the generated potential to a potential level of 230,000 volts to be transmitted over power transmission lines at the plant's main switch yard and then carried out for use by the consumer customer at his plant.

These generators are a type of synchronous generator, which was described earlier with regard to the parts it uses, such as the rotating part (rotor) and the fixed part (stator), in addition to the exciter part.

It has been described earlier that the Hoover Dam hydro power plant is considered the largest hydro plant in the USA, which generates power at 16,500 volts potential level. The potential level is sent through the main power transformer, shown typically below, which raises the generated potential to a potential level of 230,000 volts to be transmitted over power transmission lines at the plant's main switch yard then the generated energy is carried out for use by the consumer customer at the consumer plant.

It has been described earlier that the Hoover Dam hydro power plant is considered the largest hydro plant in the USA, which generates power at 16,500 volts potential level. The potential level is sent through the main power transformer, which raises the generated potential to a potential level of 230,000 volts to be transmitted over power transmission lines at the plant's main switch yard then carried out for use by the consumer customer at his plant.

The Hoover Dam hydro-turbine-driven power generation plant has 17 units of turbine-generator units. Each unit generates approximately 133 megawatts of power to make the total generated electric energy of the hydro plant nearly equal to 2,070 megawatts by the hydro plant (power transmission lines) at the plant's main switch yard then carried out for use by the consumer customer at his plant.

3-2-3. Hydro Plant Power Supply Panels

Power supply panels at the hydro plant are not different than any panels as far as powering the auxiliary motor loads that drive mechanical loads. The power supply potential varies with respect to the different sizes of the driving motors, which operate the auxiliary mechanical load.

Like any other power supply panels they are constructed with vertical as well horizontal electric bars connected with each other, and provided with the required air circuit breakers, in addition to power, and control wires to furnish the required power to its associated loads, which can be motor loads or other related loads such as connected pumps, or other electric loads.

The motor load usually is connected to its associated circuit breakers, which can be switched on or off by associated control switches mounted on the surface sides of the power supply panel in addition to control switches mounted on the main control board at the main control room per the design requirement of the hydro plant; however, this type of control (control room) is associated with big motors, or general control requirement, and such requirement will be dictated by the instrument control engineer, which also requires client instruction and approval.

Certain main power supply panels require protective relays instruments, which are installed in the main power supply (panel structure), in addition to indicating instruments as required by the design criteria. Such requirements were addressed earlier, and it will be elaborated upon more later in this write-up.

3-2-4. Electric Auxiliary Equipment

There is always a need to include the requirement of electric auxiliary equipment in every power generation plant, and as such the hydro plant is not any different from other power generation plant. The hydro power plant requires the use of battery-banks of DC power source to operate certain electric load, including powering the plant instruments and control requirement in a matter similar to the operation of other power generation plants.

3-2-5. Uninterrupted Power Source (UPS) Essential Equipment

It is imperative to use uninterrupted power source in the operation of most instruments and control system. The essential power source is composed of the AC as a main power source to power a specified power distribution panel powering the associated instrument and control loads, followed by a sensor to detect the loss of the AC source, in addition to a battery-bank DC source and a DC power source (invertor) to invert the DC source into an AC source to operate all the requirement of power supply to the essential instruments and control loads without any power interruption.

3-2-6. Plant Battery-Banks (Direct Current) Power Equipment

As stated previously, battery-banks play measure role in every power generation plant, name it—hydro plant or other power generation plant. The DC power source has a generated potential rated from 125 volts to 12 volts of DC power. Battery-banks are charged with its DC potential by an AC power source provided by the power generator of the plant. A complete description of the battery-banks have been described earlier in this write-up, which can be reviewed further. It needs to be restated here again that all auxiliary DC power source equipment needs to be in accordance with the explosion proof specific requirements.

3-2-7. Plant Instruments and Control Panels

The hydro plant has control instruments and control panels like any other plant, which send control signals (control orders) to associated controlled loads in accordance with the plant operation requirements. The control signals usually go first to local control panels or the signals go to control instruments located in the main control room, then go to the electrical loads as dictated by the design requirements; however, the actual orders are sent from sensing instruments located near the electrical loads. As noted earlier, the instrument control engineer is the design engineer who prepares the engineering study as submitted by the owner of the project and specified in the project scope of work.

The hydro plant electric loads (mostly motors) also are operated and protected by control switches and protective air circuit breakers in addition to operation (indication lights). The protection instruments are usually located at power supply panels with associated control switches, and operation indication lights, and sometimes the control switches, and associated operation indicating lights are duplicated and installed on the surface of the main control board at the main control room.

3-2-8. Plant Cable Carriers

Plant cable carriers are also used in the hydro plant as in any other power generation plants and is explained earlier in this write-up. They are two types of carriers (cable trays and cable conduits).

The cable carrier usually start from power supply centers or from control panels to deliver power to associated loads.

The cables, as explained in the previous chapters, carry power supply or control signals, which are related to associate loads; and as such the cables vary in size and insulation structure in accordance with the requirement specified in cable standards and the design procedure. An elaborate description and requirement has been stated in the preceding chapters and may be referred to in order to learn more about the requirement of cables and cable carriers.

The paths of the cable and cable carriers are also described earlier in this write-up and may be referred to it for further information. The figure below shows the typical arrangement of power center (motor control center) with associated cable tray carriers and cable conduit carriers, which can be considered a typical design arrangement of power distribution center.

3-2-9. Plant Instrumentation and Control Panels

Hydro plant uses monitoring (control instruments), which send electrical signals to control the operation of its associated electrical loads in accordance with the requirement of the hydro plant operation. Those signals usually go to local control panels or go to control instruments located in the main control room:

The hydro plant also operates by using control switches, protective relays, and indicating lights, which manage and control-related mechanical loads as required by the process procedures.

The type of controls usually located on the surface of the power supply panels to manage the power supply to the same mechanical loads. Such control may be duplicated and installed on the surface of the main control board at the main control room where also several of the hydro plants control can be monitored and controlled.

3-2-10. Plant Voice Communication and Monitoring Equipment

It is imperative that plant communication and monitoring equipment be installed at any power generation plant and the hydro plant is not different. The communication equipment usually include the public address system and the voice telephone system in addition to a monitoring system. Those equipment are very similar to the equipment required in any power generation plants. The monitoring equipment can be an annunciator system usually mounted on the surface of the main control board at the main control room in addition to a monitoring system, which is a television set. As stated earlier, most hydro plants usually use what is called a mimic bus showing in a diagram the flow of operation in the hydro plant.

3-2-11. Plant Fire Protection System

Most of power generation plants and industrial plants are required by international or local standards to have a fire protection system to assure safety of working personnel and safe operation of erected equipment. Not only that but insurance companies demand such industrial facilities should adhere to the requirement of applying safety standards during the design period of the facilities in order for the owner to be able to obtain insurance protection policy for the facility to operate within the local regulation requirements.

The fire protection equipment can be the in the form of liquid gas or in the form of foam, which is capable to stop any fire accident extension, and hence prevent the harm to reach the operating personnel and to protect working equipment, saving any cost may get involved because of the fire accident.

3-2-12. Plant Emergency Exit Signs Requirement

The emergency exit sign requirement to be installed in the hydro plant is as important as the requirement of having fire protection equipment. The signs are usually located near closed offices as well as in critical areas of the plant.

Such equipment operation are equipped with self-operated batteries as well as using an uninterrupted AC power source for the safe exit and use by operating personnel of the plant and public alike to lead the exiting mass of people out of the endangered area as safely as possible.

The exit signs in any facilities are part of the safe evacuation during any emergency condition in any industrial plants including any power generation plant.

3-2-13. Plant Main Control Room

Hydro power generation plant main control room as explained earlier in other power generation plant is considered the managing and control brain of the hydro plant using a main control board to direct and manage several units of hydro plants. The main control board in the hydro plant is slightly different because it manages and controls several hydro plants. The board has several control instruments, several control switches, and operation indicating lights, with main annunciator system on one main control board.

The main control room has also a dispatch control board to dispatch power to different areas of consumer areas to receive power source as subscribed by different consumers. Main control room with main control board are shown earlier.

II. NUCLEAR POWER GENERATION PLANTS GENERAL PREVIEW

Nuclear electric power generation plant operates with the use of steam, which drive its associated steam turbine like other steam-turbine-driven power generation plants. In the conventional steam-turbine-driven plants the water boiler generates the required steam while in the nuclear power generation plants the nuclear reactor in its types variety generates the required steam to drive the power generator turbine. The auxiliary mechanical equipment in both the nuclear plant and in the conventional plant are in general similar to each other in their structures and operation except for the equipment in the nuclear plant, which uses the boiling-water reactor type, which is different because of the nature of the generated steam, which is a radioactive type of steam.

It is imperative that the engineer who prepares any studies related to any nuclear energy project must have the experience and know-how to deal with the design and construction of any nuclear project in order that his design is safe in every meaning of the word.

The effort of this write-up is aimed at in general to educate the young engineers more about nuclear energy when they should get involved in any nuclear project design activity, and related study. The detailed study will enable educated engineers be familiar with the nuclear fission process, which ultimately produce the thermal energy to produce the steam, and hence produce the working power to drive the steam-driven turbine to power the main generator to produce electric energy.

The content of this write-up will also discuss the requirement and application of international standards that the design of any nuclear facilities must adhere to when a nuclear project is planned to take place. The write-up also will display needed information to cover different types of nuclear reactors used nowadays and will be used for generating thermal energy (steam) delivered by nuclear reactors in many of nuclear power generation plants. Also to be addressed here are the descriptions of all electrical and mechanical equipment, which are used in such nuclear plants. The nuclear power generation plants in principle are not too much different from the conventional power generation plants. The nuclear reactor acts as a boiler to generate the driving steam to move the associated turbine, which is tied up to the main power generator. Furthermore, there are also in the turbine area several of lined-up turbines (high-pressure turbine, middle-pressure turbine, and the low-pressure turbine in addition to the main condenser.

The main circulating water pump, cooling water condenser, and cooling tower are the same case like in the conventional power generation plants. However, such mechanical equipment may differ in accordance with the type of reactor used in the nuclear power generation plant. The nuclear plant usually has a very distinctive landmark, namely the big building of the reactor containment structure, which contains the nuclear reactor body, steam generator, cooling equipment, and other auxiliary equipment. The containment structure is built with a steel reinforced concrete type of building, which has a wall thickness of more than 3 meters thick in order to avoid any nuclear radiation leakage, such as gamma rays to the outer side environment. The turbine assembly as described earlier is also tied up to the main power generator, which is connected to the power transmission lines through associated main oil circuit breakers.

Inside a Nuclear Power Plant

Figure shows a typical arrangement of equipment used in the nuclear power generation plant using a pressurized water reactor type.

A Containment Structure		**G** Generator	
B Control Rods		**H** Turbine	
C Reactor		**I** Cooling Water Condenser	
D Steam Generator		**J** Cooling Tower	
E Steam Line		**K** Fuel Rods	
F Pump		**L** Transformer	

1. Nuclear Energy Study

Human creature had started at early stage of his life with primitive means, and till the present time, and continued to follow up with the discovery of metal elements structure buried under earth. As part of this man inquisition he wanted to find something new that could benefit him, and the benefit of his race. The human with his ability of learning was, and still is able to use his brain for the benefit of his race to make the world in several educational media a better world not only for himself but also for his fellow man.

The honorable human effort was a good pace in the right direction; however, some of his inventions fell in the hands of some bad humans who used the good discovery/ invention in the wrong way and caused the destruction of several living societies.

Human creature with good intention started to be a science man to discover many good things to benefit his surrounding social life. One of the science man efforts was the discovery of the atomic energy, which could benefit the man kind in every aspect of his life.

The science man had learned that the atom structure in most metal elements of our world resembles to a great extent the biological atomic structure of man, and when the atomic energy is used in the right direction, it will lead him to live a better life. The benefits can be in the form of human biological cure or in other beneficial events to enhance a good survival and good life for many years to come.

The science man with such discovery and accomplishment as was described above had made him fulfill his honorable scientific message to all. However, if the science man message was misused in any way, then the science man would have been disappointed and had lost a lot of hope and good will.

The good science man continued his good effort and discoveries in order to help his surrounding community to live in a better world. This good effort led the science man to learn more about the structure of the atom, and associated nucleus, which carried floating particles inside of it, a characteristic of all metal elements' structure.

The science man stressed especially the fact about the energy of those floating particles inside of the nucleus, and how it may be useful for the benefit of mankind.

Despite the fact that the early discovery date of the atom structure could not exactly be determined, the discovery of the x-rays by the scientist Roentgen in the year 1895 had lightened the path to the start of greater discoveries and inventions, which led the scientist Einstein in the year of 1905 to say then that there is a relation between the energy in the atom and the mass of the atom.

The scientific studies later in time had helped scientist Chadwick in the year of 1932 to discover the neutron inside of the atom nucleus. The discovery may be called the key to open the door for more knowledge about the nuclear energy.

The atom structure discovery had continued its pace when both the scientist Hahn and the scientist Strassman continued their work, which was supported by the scientist Chadwick's discovery. The fission process, which had been discovered by both Hahn and Strassman later in the year of 1939 had confirmed the process was able to release energy locked in the structure of the heavy atoms of uranium. However, the scientist Fermi and his associates showed in the year of 1942 how to fit "The key into the lock," so to speak, to produce a chain reaction to uranium fission, and to take the nucleus out of the laboratory for full scale of energy release. Unfortunately, the entire process described above had announced later the production of a bomb called nuclear bomb, although it was not meant to be, and the bomb was used for the first time in the year of 1945 to destroy a whole Japanese community! Accelerated scientific activities with respect to the safe use of the discovered new atomic energy were later pursued and harnessed (uranium fission) process to produce practical electric energy. One of the industrial firms, namely the Electric Utility Firm in the United States, which is an important basic industry, used the atomic energy for the best advantage to a better future. Such effort encouraged the use of atomic energy to form a steam-generating equipment where the energy can produce the generated steam to drive certain types of steam turbine, and consequently, to drive the attached generator to produce electric energy, while other firms used the atomic energy to develop, unfortunately, atomic bombs. Humans' love of life has encouraged him to pursue a safe world instead of a destroyed world! The matter resulted into signing peace treaties among the countries who possessed nuclear bombs in such a way to prevent the use of any nuclear weapon such as the nuclear bombs or other bombs in any nuclear war especially when two such opponents possess nuclear bombs. It has been determined that such use of the nuclear weapon may destroy both of the opponents, and perhaps the rest of the world!

It is imperative to know the above conclusion does not mean that every country must have a nuclear bomb; on the contrary the aspect here is to prevent such attitudes for the keen and safety of the world.

2. Nuclear Physics

The study of nuclear physics helps the young engineer to be familiar with the structure of the main substance of uranium metal which is the subject of this section. Uranium material, when used within the giddiness of international standards, can be a good resource of unlimited nuclear energy.

To understand uranium metal atomic structure, one can use lead metal as a referenced structure to uranium metal, which is considered a referenced metal in order to analyze and to study the atomic structure of uranium, and as such if one looks at the outside shape of the lead metal, it is found to be a soft metal, which can be formed, to a certain extent, into any shape; and no matter what is done to lead metal it will stay a lead metal. Now if the lead metal was looked at under a sensitive microscope, it is found the lead metal consist of many crystals lined up with each other very casually. Now again if one of the crystals was looked at more closely, it is found that each crystal consists of small particles lined up uniformly side by side and called one of those particles an atom. It was found that one million billion of those atoms may not cover the head of an average small pin.

Scientists continued their research looking inside the space of the atom, which appeared to be a vast space relatively speaking similar to the vast space of the universe, and found out in the center of the vast space of the atom an extremely small particle called nucleus.

The scientists found also other smaller particles that circle uniformly along the circumference of the atom called electrons, and those electrons circle their path in a very vast speed, and in a path equal to the circumference of the atom:

More discoveries were continued by the scientists to look inside of the nucleus and found that there were two types of particles. The first type is called a proton, which carries a positive electric charge, and the second type is called a neutron, which does not carry any electric charge. These two particles made the scientists think that there must be some reaction to happen in the nuclear aspect as it happens in a chemical reaction; however, in the chemical reaction the process involves only the exchange of the position of electrons between the two elements reaction, and in the nuclear reaction the process belongs to the nucleus alone without involving the electrons. The scientists found further that if the nucleus was bombarded by an outsider neutron, the neutron will enter inside of the nucleus to create a kinetic excitement making the nucleus to lose one of its neutrons to produce in the process a kinetic energy, which emit a thermal energy much greater than the thermal energy emitted out of burning average coal! The process of both cases can be understood further as compared by scientists to say that the thermal energy emitted by splitting 450 grams of uranium nucleus is estimated to be equal to thermal energy emitted by burning 1,170 tons of coal, a matter that really encouraged scientists to use the great thermal energy produced out of splitting the nucleus of uranium metal because the economic benefits.

The aforementioned remarks about the thermal energy has encouraged the use of nuclear energy when it comes to electric power generation because of the vast economical savings.

It is so imperative when using nuclear energy to aim at its normal safety requirement by adhering to observe all safety requirements, which limit any nuclear radiation harm cause. This emphasis is amplified through learning more about three types of nuclear radiation emission, namely alpha radiation, beta radiation, and gamma radiation, which the latter is the most harmful type of radiation.

Knowing now some of the dangerous aspect of the aforementioned radioactive rays, especially gamma rays, which has electromagnetic effect similar to x-rays and microwave rays, it is very important to have the required protection dealing with those dangerous rays.

3. Nuclear Fission

Natural uranium metal is considered an important material in the nuclear fission process. The uranium material in general has three isotopes, namely Isotope U-238, which makes 99.3% of the metal material, Isotope U-235, which makes 0.7% of the metal material, and lastly Isotope U-234, which makes extremely little of the metal material.

However, Isotope U-235 is considered the most basic part of the metal material in the nuclear energy phenomena.

The scientists found that either of the fast or the slow neutrons are capable to fission the nucleus of the Isotope U-235, while only the fast neutrons are able to fission the nucleus of Isotope U-238). It was also found that when the nucleus of Isotope U-235 is bombarded by a neutron, it captures the neutron and form a nuclear compound full of a massive energy, which makes the compound unstable to release immediately several neutrons.

The nuclear fission process and the release of neutrons is considered an estimate process to give an average count of the released neutrons as the result of nuclear fission, and it cannot be forecasted exactly what may happen in a certain way of nuclear fission process. It can be said also the fission process may effect Isotope U-238 when one of the released neutrons due to fission Isotope U-235 is picked up by the Isotope U-238 to produce unstable nuclear compound, and the compound produces gamma rays, and the compound may dissolved to produce a new Isotope called U-239, and the process continues on to produce another Isotope (P-239), which can be fission by fast neutron or by slow neutron emitted by the Isotope U-235, a process which is used in developing a nuclear bomb. The entire nuclear process is shown in the figure below:

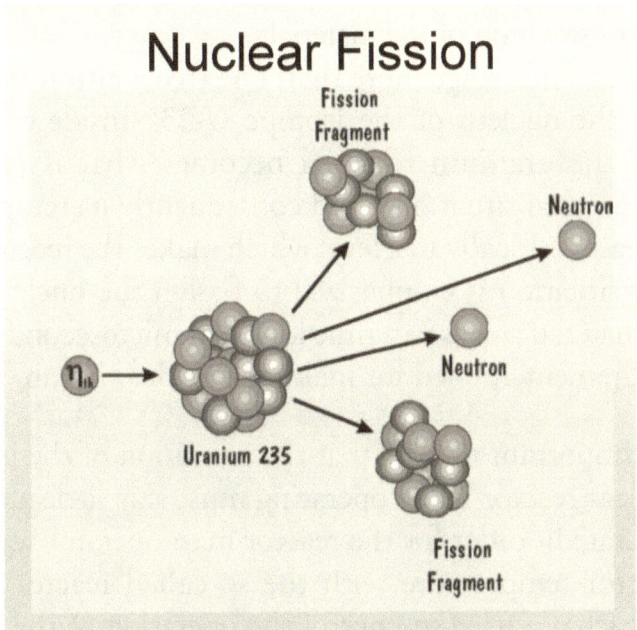

Nuclear Fission Process

The proceeded description of nuclear fission process is similar to the fission process, which takes place inside of a nuclear reactor where such fission process makes the nuclear reactor go into three situation or three alternatives, so to speak, but only one alternative could happen:

- Assuming one of the free released neutron collided with the nucleus of the Isotope U-235 inside of the reactor to split the nucleus and makes the uranium material inside of the reactor to be in a critical state, which is the suitable situation to operate the nuclear reactor.
- Assume less than one free released neutron was able to collide with the nucleus of the Isotope U-235 inside of the reactor then the uranium material becomes what is called in subcritical situation, and therefore the uranium material will need to go through enticed fission procedure to end the present situation of the material.
- Assume an average more than one free neutron to collide with the nucleus of the Isotope U-235 inside of reactor then the uranium material becomes what is called in supercritical situation), and consequently its temperature increases critically to a level which makes the reactor melt. This situation is emphasized to fission the nucleus of all the material atoms in a time less than microsecond, a must requirement applied for making a nuclear bomb.

It is important to note that the condition of the inside of the nuclear reactor while operating must stay little above the critical state in order for the reactor man operator to control the reactor temperature with the so-called reactor control rods, which are used to control the operation of the reactor, and the reactor situation stays in the state of critical level.

4. Nuclear Chain Reaction

Most power generation companies nowadays use nuclear power generation facilities, which requires a continuous use of nuclear materials fission inside of the nuclear reactor to generate required hot steam to drive the turbine-generator assembly. The process requires a continuous nuclear fission, which scientists call a nuclear chain reaction, in order to generate the required driving steam power.

5. Nuclear Radiation

The nuclear fission process emits certain nuclear radiations, and those radiation rays can be very harmful to human beings if he/she was exposed to their harm, as such engineers who work on the design and construction of any nuclear facility must have the required experience in that field, and must be familiar with the nuclear facility applicable standards to comply with. Those standards are forced by the local Nuclear Commission in any country.

The nuclear radiation rays are divided into two types; the first type is called ionic rays, and the second type is called excitation rays. The ionic rays are considered more harmful than the second type to deal with because the ability of the ionic rays to ionize most everything which the rays pass through. It is a fact to know that ionization process is separating atom into charged ions which could have been saved if the atom inside of any biological body was not exposed to nuclear radiation. The ionization process creates unorganized state in the biological cells when cells are exposed to nuclear radiation effect, which makes the natural cell in anybody to act in a crazy way, and as such cannot perform its natural duties. This condition eventually develops various cancer situation types. It is worth it here to describe the types of nuclear radiation harms, which are two types. First harm is called direct harm, which occurs when the nuclear particles collide with the biological cell and split it into two halves, and this type of nuclear particles often targets the (DNA) agent, which is inside of the chromosome of the biological cell. The chromosomes usually carry the biological map of human biological inheritance. The effected biological cells then at this stage of harm is designated a cancerous cell, which will not be able to carry on its duties as a normal biological cell. The second type of nuclear radiation harm, which is called indirect harm, usually targets the water material inside of the net of the biological cells, and if this harm was able to reach the (DNA) agent, and started to react with it, then this type of nuclear harm produces the same cancerous harm as the first type. The ionic rays usually move very fast as fast as the light, while the excitation rays move very slow and may be blocked the use of chemical protection. This discussion should be very important to learn by any engineer may get involved in any nuclear study.

6. Nuclear Power Generation Plant Mechanical Equipment

The power generation plant mechanical equipment consist of the nuclear reactor, which resembles the boiler, and associated auxiliary equipment, and several equipment that are used in conventional power generation plants. It is very important to learn more about the mechanical equipment and especially the nuclear reactor, nuclear fuel, and other equipment related to the operation of the nuclear power generation plant, particularly about operating the control rods used inside the reactor's body. The design engineer must learn also about the nuclear waste management, and how nuclear radiation affects human life. Furthermore, the design engineer must be familiar with all nuclear standards application requirements and how to deal with the safety requirement when using nuclear energy.

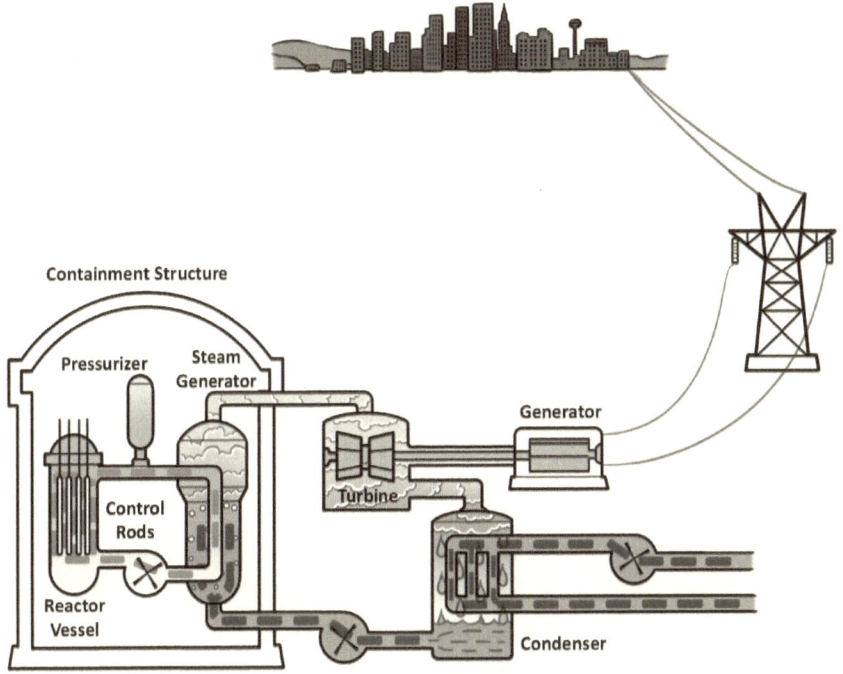

**Figurative pressurized water reactor
power generation plant**

7.   Protection against nuclear radiation

Protection against nuclear radiation is the most important aspect for the design engineer to be concerned with in his design procedure. Without this safety requirement application most Atomic Energy Commission will not approve any design, construction, and operation of any nuclear facility that does not comply strictly with the commission's applicable standards.

The applicable commission standards are aimed to provide full safety measures against nuclear radiation which time, place, distance, and quantity of bad radiation play a role in causing nuclear harm to humans.

8.   Nuclear reactors

The nuclear reactor is a special mechanical equipment similar to a boiler to boil water to produce steam. The needed heat to boil the reactor water is derived from the result of nuclear fission of the nucleus in the uranium metal atom.

8-1. Nuclear reactor types

The greatest events of the nuclear reactor invention and its variety took a place in the twentieth century. However, the most important type of nuclear reactors was the thermal reactor, which was used to generate thermal steam energy to operate a steam turbine driven to drive its associated power generator to generate almost unlimited electric energy and proved to be an ample quantity of electric power to be used in almost every aspect of life extended to serve many industrial facilities, home lighting, street lighting, and many other services. The invention helped the increasing population of the world to relax, and not to worry about the lack of electric energy, or the lack of conventional fuel, which operated a conventional power generation plants. The production of electric energy by using nuclear energy became the cornerstone in improving several aspects of life including the liberation from using imported conventional fuel.

Most electric energy utility companies use nowadays the thermal nuclear reactor The nuclear reactor has also next a thermal protective sheath called thermal shield, which keep some nuclear radiation rays, such as alpha, beta, and gamma particles rays and other emitted neutrons contained inside the reactor vessel in order to protect against any harms. The reactor vessel has also what is called reflector surrounding everything contained inside the reactor vessel in order to reflect and send back any reaction inside the vessel, and hence increases the reactor operation Efficiency.

One of the major components of the nuclear thermal reactor is called control rods, which absorb the emitted excessive neutrons inside of the reactor when needed by moving the control rods in and out of the nuclear reactor in a way to keep the nuclear chain reaction in a continuous action as required by the nuclear operation to generate the required steam, which drives the power generator. The nuclear reactor has also an important equipment called cooling system, which maintain the reactor core, and associated system at a certain temperature level to be used to heat and convert the liquid in the heat exchanger to steam sent to the associated to turbine, which drives the power generator . The operation of cooling the reactor core also helps to prevent the melting situation of the reactor if the temperature of the core passed the safe limit set by the processed procedure. It is also worth it to know that the described process of generating operating steam differ in one type of nuclear reactor from other types, which will be discussed later.

Cooling system materials also vary in one type of reactor from other types. Cooling air, hydrogen, helium, melted metals (lead-bismuth, and melted sodium are other types of coolant in addition to water are used to cool the core of the nuclear reactors.

The inside of the nuclear reactor, which is called the reactor core, consists of what is called nuclear fuel rods made of nuclear reactive material with fissionable nucleus, and is diluted with none fissionable material for nuclear causes.

The rods also can be manufactured in special fertile fuel to be ready as nuclear fuel to be used inside of the reactor. The nuclear fissionable fuel, which is used in the nuclear reactor, can be one of the following known three materials, namely U-235 or P-239 or U-233. The U-233 is the result of a nucleus of material (Thorium -232) to have picked up a neutron, then after a while this material changed to become uranium (U-233) material.

It is known that all of the above-described nuclear fission materials are metal materials, which have a defined chemical properties to be manufactured in different shapes, such as pipes or sheets or in shape of rods or in shape of small balls, or it can be changed into powders to be dissolved in certain liquids then it is used (any of the materials) as nuclear fuel in the operation of the nuclear reactors.

8-2. Control of the Nuclear Reactor

The control of the reactor means the control of nuclear chain reaction, which takes action inside of the reactor, and that requires a lot of caution not let any nuclear radiation to go out of the reactor. Again in order for the nuclear chain reaction to continue, it will need another nuclear fission process. The level of the produced thermal energy actually depends on what is called neutron density, which also depends on the number of neutrons in a volume unit called neutron flux, which is equal to the number of neutrons pass in defined area unit during defined time unit.

There are two types of neutrons that are emitted during the process of nuclear fission inside of the nuclear reactor. The first type is called prompt neutrons, which are emitted in a very fast time (less than a microsecond). The second type of emitted neutrons are called delayed neutrons. Since the prompt neutrons are so difficult to control, the control instrumentation of the reactor is designed to control the delayed neutrons, knowing the fact that the delayed neutrons take few minutes to be emitted after the occurrence of the nuclear fission process.

The reactor control depends usually on a factor called multiplication factor K of emitted neutrons, which equals to the ration of the number of emitted neutrons now to the number of neurons emitted in one previous process.

Should the ratio equal one, then the nuclear chain fission process continues, and the reactor operation stays normal to produce the required thermal energy. If the described ratio goes above one (K > 1), then the nuclear reactor operation goes above normal, and the reactor thermal energy exceeds the nuclear safe operation and causes a lot of problems.

The nuclear reactor usually stays somehow active during a limited period even it was shut down. The reason is the fact because of the related influence of the rays decay and the influence of delayed neutrons inside of the reactor take time to disappear; therefore, the coolant materials (can be water) should get rid of the heat dissipated by the process.

It is also known that the operation personnel of the reactor are compelled to stop the operation of the reactor in emergency situations control, and to transfer the emitted heat to safe places to avoid the meltdown of the entire body of the nuclear reactor.

As stated earlier the reactor control operation is done by the control rods, and in the normal operation of the reactor the control rods are immersed completely inside of the reactor core; however, the operation itself has two folds. The first fold deals with any emergency shutdown case of the reactor, which is called scrammed reactor, to immerse the control rods immediately, and completely inside the core of the reactor, while the other fold of the reactor control deals with the normal case of the reactor control to maintain a continuous a normal constant thermal energy production.

In the normal operation of the nuclear reactor, the specialized operator pulls the regulating control rods completely out of the nuclear reactor core. The indication instrument of the reactor operation tells the operator the position of the rods at that instance, and other indication instrument point the extent of the neutron density, and later other instrument tells the operator the time the neutron density took to reach the density.

The total indication of all those instruments tell that the nuclear reactor has reached the normal operating situation of "Reactor goes critical" (neutron emitted equal neutron absorbed), which means the nuclear fission chain process in the reactor has reached the normal operation of the reactor to produce normal thermal energy.

The specialized operator can also manipulate the K factor as described earlier by using the control rods to bring it to K = 1 to maintain a critical position, and hence to protect the safe operation of the nuclear reactor.

It is imperative to say that in order for the reactor to operate safely there should be a dependent continuous electric power source to supply the nuclear reactor required auxiliary equipment. This special power source is called *engineering safety feature system for reactor auxiliary equipment operation.*

The nuclear reactor needs, as described earlier, a cooling system when the reactor has what is called reactor accident to prevent the melting of the reactor should its temperature exceed the limit. The melting situation could also be controlled by immersing the control rods immediately inside of the reactor core. That operation is called scram. If this operation fails then the nuclear reactor will melt, and the nuclear chain reaction will stop; however, the reactor may emit harmful nuclear radiation inside the nuclear containment building, which should be contained inside the building to prevent such radiation to spill outside of the building.

8-3. Types of Nuclear Reactors Used in Nuclear Power Generation Plants

The market of nuclear reactors produced several types of nuclear reactors for generating thermal energy, and those reactors are called thermal reactors, which use what is called moderators to slow down the speed of very fast emitted neutrons to a speed level to nuclear fission material like U-235 and associates.

This type of reactors usually uses natural uranium material or uranium enriched little with uranium (U-235) for fuel. The reactor also uses moderator material (graphite) to moderate the reactor activity, and heavy water (H2O2) to cool the reactor operation. This write-up will cover the description of two types of thermal reactors used by power generation plants facility as discussed below.

8-3-1. Pressurized-Water Reactors (PWR)

The PWR thermal nuclear reactor is used in most nuclear power generation plant establishments, as these establishments consider such reactor to be safe more than other reactors, namely because the net result for the reactor to generate clean steam not exposed to any nuclear radiation. The steam generated by the PWR is produced by an isolated steam generator, which is a part the of PWR despite the fact that all types of thermal nuclear reactors are subjected to abide by stringent US Atomic Energy Commission and International Atomic Energy Commission regulations and safety standards throughout the world for isolating harmful nuclear radiation.

The PWR reactor is fueled with uranium (enriched some) and cooled with light water and moderator for the nuclear fission process action inside of the reactor. This type of reactor is classified to be a nonhomogeneous thermal reactor. The nuclear fuel and the control rods of this type of reactor is called the reactor core, which is located in a container called reactor vessel. The reactor is also equipped with a coolant pump to circulate the coolant water through the reactor core, and a pressurizer vessel prevents hot water from boiling to transfer the pressured heated water (nuclear radiated water) generated by the reactor to another equipment called heat exchanger or steam generator, which is located outside the reactor vessel. The heated water acts as the heating media to heat other clean water to convert it into clean steam to drive the associated turbine, which drives the associated main power generator to produce electric energy. Despite the fact that the cooling water used in the reactor has to be very clean water, clean water may sometime corrode the pipes that it flows through; therefore, additional materials, such as zirconium, will be added to the coolant to prevent corrosion of the water passes in addition to using stainless steel materials.

**The figure above shows a typical plant
of the PWR type of reactor.**

Again as stated earlier, the operation of safety equipment must be powered by certain uninterrupted power source requirement enforced by standards and regulations issued by the NRC.

The PWR reactor startup procedure has several steps. The first step starts from the instance when the reactor is in a stop position; the coolant of the reactor is brought up to a hot standby condition when the reactor is in the critical condition as described earlier, and the temperature of the coolant reaches 532°F, which is equivalent to 278°C. The reactor usually reaches the critical stage after filling all coolant pipes, and pressurizer vessel with coolant at a low pressure condition of 60 psig, after which all bubbling condition is eliminated. The coolant temperature later on is raised by what is called electric heater to make the coolant temperature reaches approximately 450°F (233 °C), which makes the coolant temperature go down, and thereafter the reactor temperature will go up, and the pressure in the pressurizer vessel goes up to 2,250 psig, then the control rods are pulled out to make the reactor ready under control signals automatic reactor control signals to bring the reactor to the critical stage ready to make the reactor start producing steam through the steam generator vessel to be sent to the turbine, and consequently to drive the main power generator in accordance to the requirement of the electric load.

The rest of the auxiliary mechanical equipment associated with the nuclear reactor operation is similar to the conventional power plant operation that was described earlier.

The reactor shutdown process takes several steps, which brings the reactor from the hot condition to the almost cool condition. This operation is done for the sake of the required routine maintenance of the reactor equipment.

The process starts by stopping the activity of the reactor and inserting the control rods in manually all the way inside the reactor core, and later adding enough material (boron) to moderate and absorb all runaway neutrons, and this will lower the temperature of the coolant to a lower stage than it was at the startup of the reactor.

The reactor at this stage will be subjected also to some auxiliary equipment called residual heat removal (RHR) in order to get rid of whatever heat is left in the reactor to bring the reactor to a safe condition.

The unused steam left in the steam generator will be pumped to the steam condenser, and after that the pressurizer operation is stopped to operate. It should be emphasized that the cooling of the reactor should be a gradual process at a speed not to exceed 70°F (22 °C) per hour as it is required by the safety and operation of the reactor. Once the safety requirement of shutdown of the reactor was established then the pressurizer vessel is opened, and the process of removing the nuclear reactive material and waste material will be removed from the vessel, and the vessel will be refilled with clean water. As described earlier, the pressurizer vessel is part of the reactor auxiliary equipment, and as such, is located inside of the containment building, which was described earlier, and contain all reactor associated nuclear equipment. As stated above, and after filling the pressurizer with clean water, this vessel will stay open to the containment building atmosphere, and therefore, the shutdown process is considered completed and the RHR system has completed its work in a very safe way.

Having described the shutdown process, the rector is now ready to be refueled with nuclear fuel.

The process starts with making sure that the coolant temperature is made to be lower than the temperature of the reactor vessel temperature. The nuclear fuel is passed to the nuclear vessel through a special canal filled with water containing boron material. The canal is connected between the nuclear storage facility and the heart of the nuclear vessel. The water excess, after transporting the fuel and completing the fueling process, will be returned to its storage place, and the reactor vessel cover will be closed. The reactor vessel and its associated pipes are then filled with the required coolant to make the fueling process complete and is ready for a new thermal operation. Again it should not be forgotten that all auxiliary equipment, which play major role in fueling the nuclear reactor, must be supplied with uninterrupted power supply called the *engineering safety feature power system for reactor auxiliary equipment operation.* It is to be helpful also to list some of the nuclear reactor auxiliary equipment, such as steam dump system, which get rid of excessive steam production, and chemical volume) control system, which control the quality of coolant water, in addition to residual heat removal system, nuclear fuel pool system, and waste disposal system, together with other effective equipment to make the reactor operates safely.

8-3-2. Boiling-Water Reactor (BWR)

Boiling-water reactor (BWR) is considered also the most used reactor in power generation plants, as the pressurized water reactor is. The BWR reactor uses light water as coolant and moderator similar to the PWR; however, the BWR differs from the PWR because the generated steam inside of the BWR is passed directly to drive the associated turbine, and there is no heat exchanger involved in the BWR as in the PWR; as a result some power generation plant owners think that the PWR is safer to use. However, both reactors' design requirement are subject to be approved by the Authority of the Atomic Energy Commission throughout the world.

As stated the generated steam in the (BWR) is a contaminated steam by the nuclear radiation emitted inside of the reactor vessel; therefore, manufacturers of such nuclear reactor have designed and manufactured certain protecting shields for all the equipment and pipes, which carry radiated steam in order to prevent any nuclear radiation to run outside the described equipment.

Furthermore, the manufacturers emphasized the act of placing all associated equipment of the BWR to be placed in a very protective center where such equipment can be operated remotely in order to provide safe operation of such equipment in accordance with the firm requirement of applying the standards and regulations of the Atomic Energy Commission.

The BWR reactor uses certain nuclear fuel in shapes of rods for its operation. The material of the fuel is UO2 type placed inside special pipes made of zirconium alloy material in addition to Zircagoy-2 to manufacture nuclear fuel.

The figure below shows in a diagram the typical relation among the main equipment used in the boiling-water reactor (BWR).

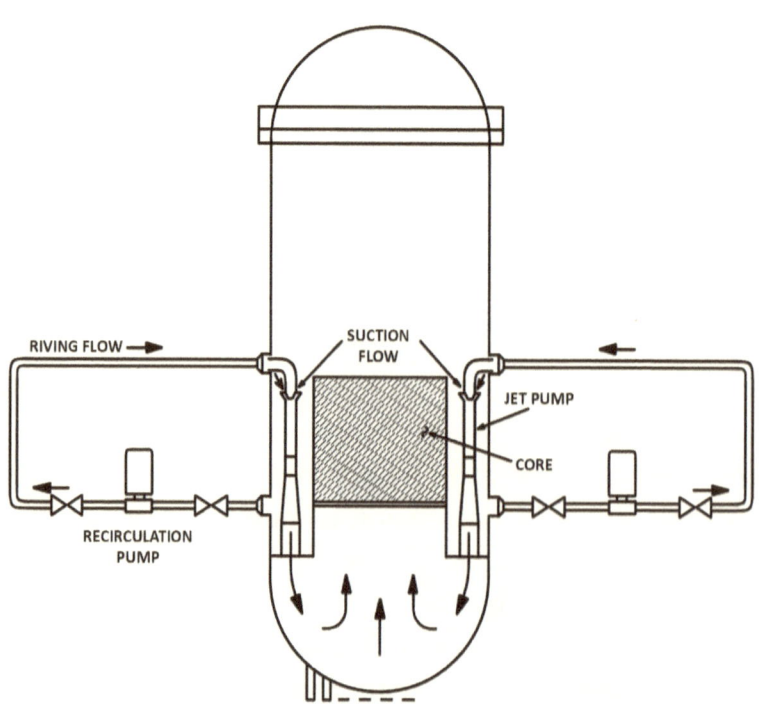

Symbolic boiling-water nuclear reactor to be contained in a reinforced concrete reactor containment building

The figure shows a reactor vessel, which is made of steel material pressured cylinder topped with a cap, and the inside surface of the cylinder is plated with stainless material to resist any corrosion that may happen inside of the cylinder. The inside of the reactor also contains the core of the reactor, which consist of the nuclear fuel, and all coolant canals, control rods, and monitoring sensitive instruments; all are surrounded with what is called core shroud, a cylinder type of shroud to isolate the coolant pipes system in the core from the outside circular partition part of the reactor. The reactor core is also connected with what is called steam separator to separate the dried steam from the wetted steam, in addition to steam dryer, which is located on top of the steam separator to dry the wetted steam, in addition to jet pumps to circulate the generated steam inside the reactor core. It is to be stated again that all materials located inside of the reactor must be made or plated of stainless steel materials.

The figure above represents a cross section of the jet pumps operation to indicate that the dried steam leaves the core of the reactor through special pipes and valves to reach the steam turbine, which moves the attached main power generator.

The entire operation of producing thermal energy inside of the nuclear reactor and processing working steam is also shown figuratively above in the figure.

The control rods system in the BWR is considered one of the beneficial safety equipment of the reactor and are located at the bottom of the reactor vessel to help by gravity to immerse the control rods fully inside of the reactor. As stated the control rods use a hydraulic operating system, and it is the specialty of General Electric Corporation design to be proud of. The control rods location has a variety of useful operation, such as not interfering with the location of the reactor (closing cap) at the top of the reactor vessel, and the fact the hydraulic system helps to control rods in moving up or down inside the reactor's normal operation or during scram function operation to stop the reactor work.

The BWR has also associated auxiliary equipment as the PWR to aid the reactor in its normal operation as well as during emergency operation especially during loss of coolant operation (automatic control), and needless to say, those equipment will be power supplied by a power source as shown in the single line diagram below:

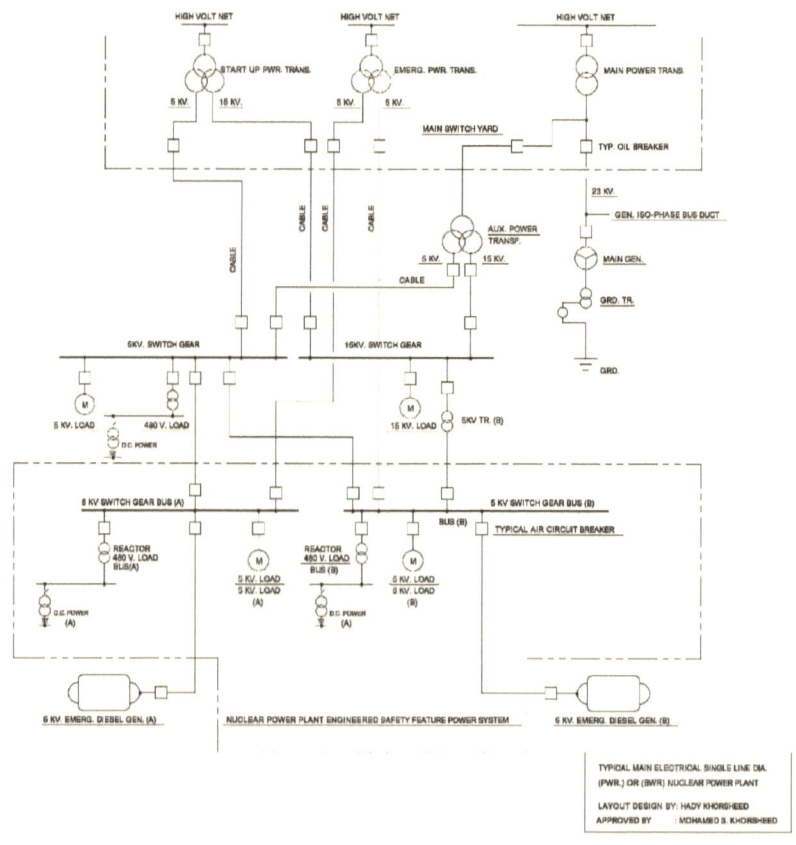

Nuclear power plant electrical single line diagram called *engineering safety feature power system for reactor auxiliary equipment operation*

The (BWR) work starts with the operation of control rods and the operation of associated coolant water pumps to circulate the coolant inside of the reactor, and such operations receive electrical signals in accordance with the requirements of the thermal energy, which corresponds to the electrical loads to be supplied by the main power generator.

The BWR may be stopped like the PWR reactor mainly by the operation of its control rods; however, this process is done with the help of the suppression pool, which able to lower the temperature of the reactor core, and consequently to stop the reactor.

The BWR is considered to be a fixed-pressure type of reactor; hence, the level of its potential is subject to the movement of the control rods, and the coolant circulation through its associated pipes, in order to provide the required steam driving the turbine, which tied to the generator to produce electric energy as demanded by the electrical load.

The BWR operation during normal work has additional auxiliary equipment as listed below:

- Reactor Water Cleanup System
- RHR Residual Heat Removal System
- Fuel Pool Cooling and Cleanup System
- Closed Cooling Water System for Reactor
- Radioactive Waste Disposal System

The reactor also needs additional equipment, which are listed below:

- Standby Liquid Control System
- Reactor Core Isolation Cooling System
- Residual Heat Removal from within Containment Building

- Automatic Steam Blowdown System
- Pressure Relief System

As indicated earlier, most safety equipment are usually redundant equipment powered by the use of engineered safety feature power system for the safe operation of the nuclear reactor.

8-3-3. Miscellaneous Nuclear Type Reactors

The nuclear reactors industry, regardless of producing the most practical and popular reactor types such as the PWR, BWR; however, the industry produced other types such as sodium-graphite reactor, which is used in smaller nuclear power generation plants, and fast-breeder reactor, which produce nuclear waste materials used in some nuclear weapons called depleted uranium weapons.

8-4. Nuclear Plant Auxiliary Mechanical Equipment

The nuclear power generation plant uses also steam turbine, which consist of three stages, (high-pressure turbine section, middle-pressure turbine section, and low-pressure turbine section) like other turbines used in conventional power generation plant. However, as stated earlier, there are two types of nuclear reactor, namely the PWR reactor and the BWR reactor; the latter reactor as stated earlier produces contaminated steam with nuclear radiation, which drives directly its associated turbine, and as such the turbine must be equipped with protective shield in accordance with safe regulation standards requirements to prevent the contaminated steam to exit out and produce a hazardous condition within the plant.

The nuclear reactor has also other auxiliary equipment, such as the steam condenser, which is also equipped with a protective shield as listed above, in addition to the coolant pumps, and RHR equipment to do the heat removal after the reactor being stopped, and the chemical and boron injection to treat the used coolant water, and other required equipment, which was listed earlier. Again needs to remind that all safety equipment must be powered by a power system shown above called *engineering safety feature power system for reactor auxiliary equipment operation.*

9. Nuclear Reactor Electrical Equipment

9-1. Main Power Generator

The main power generator is as always considered one of the main equipment in the nuclear power generation plant and is mounted on a seismic qualified turbine-generator pedestal, which should withstand any kind of vibration including earthquake effect, and the sole vibration of the generator-turbine assembly without inducing any effect on the assembly normal power generation operation. It should be also emphasized that the BWR reactor type requires that the assembly of such equipment must be placed in an isolated area away from area which is inhabited by humans, animals, and plants because the possibility of releasing accidently any nuclear contaminated steam. It is also required that all types of PWR and BWR turbine-generator assembly and associated auxiliary equipment must be located in geographic areas which should meet the seismic qualification requirement stated by the authority of the local Atomic Energy Commission (AEC) (nuclear regulations and standards) which it will be listed later in this write-up.

The description of all electrical equipment associated with the nuclear power generation plant are very similar with the equipment described in the conventional power generation plant section except the fact the electric equipment listed in this section of this book must meet the requirement of the

AEC (nuclear regulations and standards) which will be stated later in this section.

For full description of the main power generator, and associated equipment, please refer to chapter 2.

9-2. Nuclear Plant Power Transformers

As indicated earlier, every power generation plant uses power transformers, and as such, the nuclear plant is not any different except when it comes to comply with the AEC seismic requirement standards; therefore the list shown below will identify the transformers used in the nuclear power generation plant. However, for the detailed information on each equipment, one may refer to chapter 2.

- Main power transformer, which connect the transformer with the utility main power transmission lines. This transformer is usually located at the main switch yard with associated control system. This power transformer is shown earlier.
- Auxiliary power transformer was described earlier in chapter 2 and is supplied by main power generator to supply with power source to feed auxiliary electric load.
- Startup power transformer, which is similar to the transformer described in chapter 2, and is also supplied by the adjacent power transmission lines passing by the location of the nuclear power plant.
- Emergency power transformer is an additional power transformer to provide power source to nuclear reactor associated load through the engineered safety feature power system redundant power supply panels. It is important to mention that the emergency power transformer receives its power source through power transmission lines in accordance with the standards issued by the Atomic Energy Commission (AEC), which the licensing agency

is permitting the nuclear power plant operation. This particular transformer is usually located at an independent place away from the other power transformers.

9-3. Power Supply Switch-Gears and Power Supply Panels

The power supply source in the nuclear power plant consists of two types, one type called engineered safety power system, and a system which is a redundant power system type supplying redundant power supply panels to operate the reactor electric load associated with the safe operation of the nuclear reactor. The other type of power source is called conventional power system to supply power supply panels, which powers auxiliary equipment that is not related to the safe operation of the nuclear reactor.

Power supply switch-gears and power supply panels are described in detail in chapter 2 as both types have the same structures except for the nuclear related power source equipment, which must meet requirements of the AEC regulations and standards. These power supply switch-gears and power supply panels are called engineered safety features power system as frequently is referred to in this write-up shall be the type of redundant equipment to provide electric energy to redundant electric load as required by the AEC regulations and standards; not only that but they also must pass the required seismic qualification tests.

9-3-1. Engineered Safety Feature Power System

It has been indicated and has been emphasized earlier how very important it is to have an independent power supply system to have for the safe operation of the nuclear power generation in general, and very dependent power source for the safe operation of the nuclear reactor in particular. This type of power source the Atomic Energy Commission has defined as *engineering safety feature power system for reactor auxiliary equipment operation.*

The nuclear reactor equipment normally uses the constant power source available by the auxiliary power transformer in a redundant type of power source. Should the normal power supply source be not available for any reason, then either the emergency power source will take over to power the nuclear reactor auxiliary equipment or the emergency diesel generator units will take over to power the nuclear reactor auxiliary equipment.

The AEC, the official commission of nuclear regulatory authority set certain standards, which require the use of redundant power supply to power all nuclear reactor safety equipment by redundant power supply panels in such a manner when one power supply panel fails, the other redundant power supply panel will take over to power its associated electric load. The process takes place for the operation related with the reactor safe operation.

All the power supply panels associated with the safe operation of the reactor are powered in general by the emergency power transformer; however, in any emergency case, regardless whether the emergency power transformer is available or not, the emergency directive orders will start automatically the emergency diesel generator units and bring it to stand by position but not be connected to the redundant power supply panels except when the emergency power transformer fails condition, which means that the emergency diesel generator units stay ready in operation to provide a power source should other power sources fail to operate.

Should other emergency conditions occur such as loss of coolant accident, or any accident like reactor work failure then all emergency power sources will be put on alert including the startup of the emergency-diesel generator units to make sure of the readiness to supply the required emergency power.

It is to make clear that the nuclear power plant uses all types of volt potential ranging from 15 KV down to 24 volts, which are provided by the *engineering safety feature power system for reactor auxiliary equipment operation.*

9-3-2. Conventional Power Source System

This type of power source is not involved in the nuclear reactor normal or emergency operation. It is a source to operate some auxiliary mechanical equipment to do a routine work.

The power supply panels of this type are not required to be the redundant type of panels and normally receives power from either the auxiliary power transformer or from the startup power transformer only. The detailed description of the noted panels are indicated in chapter 2 and can be reviewed as needed.

9-3-3. Essential Power Source Requirement

Every power generation plant has essential equipment as turbine axle-turning gear, turbine axle-bearing oil pump, which require direct current power supply (DC) to operate such equipment in addition to supply DC power supply to operate the air circuit breaker. This DC source is supplied by battery banks, which are connected to a DC power supply panel. The battery banks are charged by its associated charging unit connected to an AC power source, which is converted to DC source to charge the battery banks. Should the DC power source be required in the safe operation of the nuclear reactor then it will be required to the redundant equipment application.

The essential power source also requires an uninterrupted power source (UPS) system to operate the nuclear reactor required control instrumentation, which is derived from the DC source together with additional AC source. The detailed description of the UPS system is indicated in chapter 2.

However, the redundancy requirement is also required during the safe operation of the nuclear reactor.

9-4. Cable Trays and Cable Conduits Requirement

As stated earlier in all power generation plants there are two types of cable carriers, namely the metallic cable trays carriers and the conduit cable carriers, and all these carriers are described in chapter 2. However, all cable carriers associated with the safe operation of the nuclear reactor must comply with the AEC with regard to the redundancy requirement and Class IE equipment classification including equipment seismic qualification requirement.

The cable carriers redundant path location shall also be separated apart from the other redundant path location for safety reason caused by fire or mechanical harm as required by the AEC regulation and standards.

Figure 9-4-1. It is imperative but to state that all cable sizes and insulation must also meet certain cable selection standards in addition to the requirement of the AEC with respect to fire resistance and other requirements.

9-5. Battery-Banks and DC Power Source Plant Requirements

Battery banks providing DC power source play a major role in providing essential load power supply as indicated earlier, and this power supply can provide two levels of voltage, namely 250 volt DC for motor operation and 120 volt DC to operate air circuit breakers in addition to 24 volt DC to operate control instruments. As also stated the DC power source operate UPS system after it has been converted into AC system. The detailed description of the DC power system is detailed earlier in chapter 2. Again it is worth indicating that all DC equipment located in battery rooms must be explosion-proof type of equipment because of the nature of the gases emitted in the room.

9-6. Nuclear Instruments and Control Panels

The nuclear power generation plant achieves its work by the help of what is called instruments and control panels like any other plant where control instruments send electric signals to the associated electric loads to give those control orders to control the work of the plant. These orders usually go first to local control panels or to control instrument at the main control room), and consequently reaches the associated operative electric load, and such process proceeds in accordance with plant mechanical process requirements in conjunction with the operation of control switches, protective relays, and indicating instruments and lights. Such control may be located on the surface of power supply panels and duplicated on the surface of the main control board at the main control room.

Since requirement of the nuclear power plant falls under the requirements of the AEC, most of the listed equipment and associated control must meet the requirements of redundant type of equipment and control instruments specified by AEC regulations and standards and definitely the requirement of Class IE equipment requirement specifications.

9-7. Plant Communication and Monitors System

Communication monitors system, safety cameras, public address and telephone, and loud speakers requirements described in chapter 2 are very sensitive equipment. The AEC's strict standards requirements for the safe operation of any nuclear power plant operation are mandatory to be included in the engineering design of any facility using nuclear material. It is usually the instruments control engineer to plan and to include a complete engineering study of a plant communication and monitors system in the engineering design of the nuclear power generation plant. Furthermore, the same assigned engineer will be required to discuss the safe implementation of the AEC regulation and standards to be adhered to in the described designed equipment, and to establish complete approval of the design by the AEC.

9-8. Plant Fire Protection System Requirement

Fire protection system is one of the most important service requirement in the nuclear power plant to provide a full safe protection for all the employees who work at the plant, and to offer a complete safe service to the operation of the nuclear plant, which also is a requirement to establish a complete protection insurance policy issued to the plant, and without such policy, the AEC will not allow the plant to operate! For a detailed description of the fire protection system, refer to chapter 2.

9-9. Emergency Exit Signs Indication System Requirement

Exit signs requirement is one of the AEC requirement to comply with for helping plant personnel and other people inside of the plant to get out safely of any troubled area in the plant. It is also a requirement dictated by the plant insurance policy. For a detailed description, refer to chapter 2.

9-10. Nuclear Plant Security System Requirement

The nature of this type of plant having nuclear material and nuclear fission operation require top security system which include security operation personnel, and a security instrumentation to alert all people working in the plant and people who visiting the plant to alarm the people about the safe present of all people should a wrong behavior of any nuclear accident may had happen inside the plant may necessitate a major inspection or evacuation of the plan.

9-11. Nuclear Plant Main Control Room

Main control room is considered the directive brain for all the equipment in the nuclear power plant. The main control board in the main control room has control instrumentation, control switches, indicating instruments, annunciator board, and main mimic bus.

The main difference between this board, and the board described in chapter 2, is the fact all instrumentation requirement must be the redundant type as far as the nuclear board section and its associated auxiliary equipment is required.

Furthermore, the main control board has a separate section especially for the control of the nuclear reactor and associated equipment.

The main control board and associated equipment must also meet the Class-IE equipment requirement and equipment seismic qualification as required by AEC standards and regulations. The AEC requires also installing monitoring equipment throughout the nuclear plant to detect any nuclear radiation spread out in order to alarm employees in the control room about the hazard and the safety of the entire plant, and consequently to control the hazard.

10. Plant Compliance with AEC Applicable Regulations and Standards

The nuclear power generation plant is different from the conventional power plant when it comes to comply with AEC regulations because of the strict compliance with the safe operation requirement of the nuclear plant.

Those standards are also updated frequently to place a better safety requirement on the nuclear power plant safe operation, and last but not the least when the nuclear plant goes in operation, it must comply with the requirement of quality assurance criteria for nuclear power plants identifying the nuclear engineering study in accordance with all nuclear regulations and codes (NRC). This agency was founded in the United States to issue some scientific standards for applying the safe requirement of the quality assurance document and is listed in the Code of Federal Regulations, Title 10, Part 20 in the Standards for Protection Against Radiation and Part 50, Licensing of Nuclear Production Standards. This last part (50) contains two parts (10 CFR 50 Appendix-A), which discuss the nuclear plant description and design and part

10 CFR 50 Appendix-B, specifies the disciplinary action of the type of equipment to be used in the nuclear plant. An additional AEC regulation (Part 55, Operators Licenses) requiring training employees of the nuclear plant to operate safely the starting and shutdown the nuclear reactor and associated equipment, and Part 71, Packing of Radioactive for Transporting, for transporting nuclear waste.

Part 100, Nuclear Power Plant Site Criteria, states requirement to select general location for a desired nuclear power plant location, and Part 140, Financial Insurance Protection Requirements, is to select nuclear plant insurance coverage, and finally, Part 170, Fees for Facilities and Materials, requirement to renew plant operating license.

11. Plant Adherence to Application of Nuclear Regulations and Standards

The scientific regulation study, and professional experience have helped very well in the study of nuclear energy and other studies, and as such the (AEC) has adopted those regulations in most of its requirement when it comes to use those regulations in the application of the design nuclear reactor facilities, to force those regulation to be adhered to when it comes to design nuclear power generation plants; therefore, some those regulations will be listed here under together with the institution who formulated the applicable standards.

11.1. Institute of Electrical and Electronic Engineers

The institute has many publications that are adopted by the AEC to guide its application and be adhered to when it comes to the electric system in the nuclear plant as listed below:

* IEEE Standard-279 is applied and be adhered to in the specification of nuclear reactor protection system.
* IEEE Standard-308 is applied and be adhered to in the specification of auxiliary electric equipment, which is classified as Class IE equipment.
* IEEE Standard-317 is applied and be adhered to in the specification of electric power supply, which passes through reactor containment building.
* IEEE Standard-323 is applied and adhered to in the specification of electrical equipment (Class-IE equipment), which withstands the minimum effect of earthquake and other environment effects.
* IEEE Standard-326 is applied and adhered to in the specification of standby diesel generator unit (emergency equipment), which can withstand any interruption to provide adequate power for the operation of nuclear reactor.
* IEEE Standard-336 is applied and adhered to in the specification of control instrumentation, which is used with nuclear reactor plant during the construction of nuclear plant.

- IEEE Standard-338 is applied and adhered to in the specification of routine test of protection equipment in the nuclear plant, and to maintain records of such tests.
- IEEE Standard-340 is applied and adhered to, to make sure of the readiness of emergency diesel generator power available in emergency conditions.
- IEEE Standard-334 is applied, and adhered to in the specification of electrical equipment including specification of Electric Wires) used in the nuclear power plant.

There are also additional standards that are applied and adhered to in the design and construction of a nuclear power generation plant, which will be discussed later in this write-up.

11-2. References of other Scientific Studies Associations

11-2-1. American Society of Mechanical Engineers (ASME)

The American Society of Mechanical Engineers of the major society in the United States has prepared and printed several standards about the requirement of the mechanical equipment used in all power generation plants, some of which are listed below:

- Boiler and Pressure Vessels Requirements
- Turbine Design Requirement
- Mechanical Flexibility Analysis Requirements

The above-listed standards are required to be adhered to by the AEC when it comes to preparing preliminary safety analysis reports (PSAR) and the final safety analysis report (FSAR) for the nuclear power generation plant engineering study.

11-2-2. American National Society of Instruments (ANSI)

The American National Society of Instruments (ANSI) has prepared and printed several standards to guide any engineering studies, especially when it comes to prepare a study for nuclear engineering study especially the safe operation of the nuclear power generation plants during geological effects (earthquake) and other effects, and during harmful effects of environmental accidents such as flooding and other effects.

11-2-3. American Society of Monitoring Instruments (ASI)

This association has prepared certain specification for all types of instruments used in the nuclear power generation plants and other plants for the safe monitoring of such plants and other plants also qualifying such instruments to withstand the effects of earthquakes.

11-2-4. Insulated Power Cables Engineering Association (IPCEA)

The documents of this association is specialized in the requirement and specification of electric cables insulation, and other requirement for the use in nuclear power generation plants as required by AEC especially when it comes to comply with the requirement of to withstand the effects of fire and the aging of power cables and passing all the tests in accordance with the AEC regulations and standards in addition to test requirements specified in DEMA and NEMA standards for the use in nuclear power generation plants.

12. Radiated Material Waste Contamination Effects

Although nuclear power generation plants are considered to be clean plants, nevertheless the nuclear fuel gets to be consumed and produces nuclear waste. However, the US government and the AEC has put out extreme laws and regulations to treat and get rid of such harmful nuclear materials. The most harmful waste materials are listed below:

- The nuclear-radiated waste, which is the result of fission.
- The nuclear-radiated waste, which is the result of the neutron captured by other material existing in the nucleus of other radioactive material.
- The nuclear-radiated waste, which is the result of the decay of the byproduct materials produced spontaneously. The produced materials in the previous item is called the nuclear aside radiation byproduct and can be classified in three types, namely:
  - Nuclear waste byproduct may exist in nuclear fuel, which is designated as Type-a and is estimated to be 99% of the nuclear waste and can stay in nuclear fuel and is depleted with the nuclear fuel exhaustion and packed in accordance of AEC requirement, then moved out of the plant premises to legally get rid of it .
  - Nuclear waste byproduct may exist in liquid used in the nuclear plant and is designated as Type-b, which will get rid of continuously during the working life of the reactor for the safety of environment, and to keep the cooling water in pure condition.
  - Nuclear waste byproduct, which may take place inside nuclear auxiliary equipment, and is designated as Type-c, may stay in the equipment during the life of the nuclear power generation plant, which is estimated to be forty years, and can get rid of it during the continuous scheduled maintenance of the nuclear plant.

It is imperative to state that the nuclear waste byproduct causes an extreme danger to the safety of humans, animals, and plants; therefore, the US Atomic Energy Commission (AEC) and similar commissions throughout the world have issued extreme regulations and standards for the work to get rid of the nuclear waste byproduct in safe manners and processes, to transfer, and bury such dangerous byproducts in places far away from humans, animals, and plants locations, in addition to planning a continuous work of keeping the plants atmosphere clear and clean!

13. Safety and Security of the Nuclear Power Generation Plant

The nuclear plant work depends on several issues, and the most important issue is the safety and security of the nuclear plant follow this issue is that the nuclear reactor capability of producing enough thermal energy to operate its associated steam turbine, which is tied the main power generator, and consequently the power generator is able to supply the connected load with the power requirement. This sequential requirement is dictated by the facts that the neutron flux is able to support the continuation of the nuclear chain reaction, which depends on the multiplication factor K as indicated earlier, which control the level of nuclear fission process and consequently to avoid increase in thermal level, which may cause melting the reactor core, and other problems such as harmful nuclear radiation emission to hurt everyone in the plant, not to mention the financial problems affecting the restoration the plant.

The AEC has made it clear that the design and operation of the reactor must adhere to the severe application of all the regulation and standards issued by the commission for the safety of the plant and the safety of the plant employees.

The AEC in every part of the world where nuclear reactor is built will not license any nuclear facility to operate till the commission make sure that the facility has adhered to the application of all required regulations, and standards issued by the commission, and that is done when the commission asks the owner of the nuclear facility to submit to the AEC what is called safeguards report covering the engineering and design of the nuclear plant in question, and the report must contain the following:

- The general picture of the plant description and its design details.
- The general picture of the plant operation procedure satisfying safe operation requirement of the plant.
- Special report describing the manners adopted for the safe operation of the plant.
- Special report describing the manners of analyzing, and study the failure accident event if it happens in the nuclear plant.
- Special report describing the manners of copping up with the accident failure within the safety requirement of the nuclear plant.

The engineering firm designing the nuclear plant facility is obligated to prepare and submit special report covering the possibility of accident failure of the nuclear plant, and the percentage of the possibility for such accident to occur in the plant.

14. Nuclear Power Generation Plant Engineering Team Qualification

Engineering experience qualification is a must when it comes to designing a nuclear power generation plant in accordance with the AEC regulations and standards requirements. The engineering team consists usually of project manager heading a group of a variety of experienced engineers specialized in different disciplines of engineering, and each discipline is headed by a group leader, and the group will carry the required study in accordance with a write-up called project scope of work prepared by the owner of the project. The engineering team is also accompanied by a quality control group to monitor the engineering work to comply with NRC requirement. The project manager is considered the main responsible engineer to see the project executed in accordance with AEC requirement in addition to the requirements of other scientific standards.

15. Client and Nuclear Commission Review of Study

15-1. First Engineering Review of Study

Every engineering discipline meets separately to discuss the requirements of the regulations and standards that his team needs to abide by and to apply its requirement pertaining to that particular team study, and for the team to mention its compliance in what is called preliminary safety analysis report (PSAR), which will be submitted to the required department in the NRC for review and approval. The engineering firm, right after it gets the approval for the PSAR, starts to prepare another required report called safety analysis report.

The NRC in the United States established typical formats that need to be followed when preparing the PSAR, SAR, and the FSAR reports. They called these formats standard formats. However, the final safety analysis report (FSAR) is considered the most important report, which the NRC reviews very carefully to make sure that the engineering study for the nuclear plant has adhered fully with its requirements. Furthermore, the engineering firm must send all engineering drawings with the FSAR for approval or to deny the approval of the mentioned documents if it sees no complete compliance with its regulations.

15-2. Content of PSAR and FSAR - Nuclear Engineering Study

It has been stated that the Nuclear Regulation Committee has established the PSAR and FSAR in order for engineering companies to follow, and those reports are the results of engineering study, which were prepared by several specialty of engineers, and later for the NRC to review and to approve of. The content of those reports will be listed in the general folder of the study, detailed as shown below:

15-2-1. Preface of the nuclear project and general description of the nuclear power generation plant.

15-2-2. Description of the general plot of land description of the geological location of the plant located away from humans, animals, plants, and other related matters.

15-2-3. Description of plant construction design and related parts, especially the adherence to the requirement for the plant to withstand the effect of earthquakes and environment effects as follows:

- Description of how the instruments and control equipment meet requirement of Seismic Design Category I- Instrument and Electrical Equipment.
- Equipment (seismic classification) requirement.
- Description of Design of Category I-Structure (Class I –type).
- Description of mechanical (dynamic tests and analysis).

15-2-4. Description of the design of the nuclear reactor and adherence to safe operation including engineered safety power system feature and control.

15-2-5. Description of the design of the coolant system of the reactor, including the associated equipment pumps, steam generator, and coolant pipes.

15-2-6. Description of all systems (engineered safety feature system), which support the safe operation of the nuclear plant in case of emergency shutdown for the purpose to contain any nuclear radiation.

15-2-7. Description of instrumentation and control system of the reactor operation surveillance.

15-2-8. Description of auxiliary power system, which consist of the following:

- Offsite Power System
- Onsite Power System

15-2-9. Description of reactor (auxiliary mechanical system), such as nuclear fuel storage and coolant system.

15-2-10. Description of steam and power conversion system.

15-2-11. Description of radioactive waste management and removal.

15-2-12. Description of the method of how to protect against radioactive radiation, which require the following:

- Radiation Protection Requirement

- Shielding Requirement
- Ventilation Requirement

15-2-12. Description of how to operate nuclear power generation (conduct of the nuclear plant operation).

15-2-13. Description of preliminary tests for nuclear plant equipment to be familiar with the equipment.

15-2-14. Description of analytical steps (accident analysis) taken to avoid accident failure of the nuclear power generation plant.

15-2-15. Description of technical methods to operate the nuclear power generation plant.

15-2-16. Describing adhered work with quality assurance for plant operation requirement. This step is considered very important for the engineering study of the nuclear power generation plant with regard to the safe operation of the nuclear plant and has been prepared and listed in the NRC published document "Quality Assurance Criteria for Nuclear Power Plants Operation Appendix-B" as noted in the 10CFR main document.

It has been made clear by the AEC that no nuclear power generation plant will be licensed to operate without documentation to assure that the nuclear plant has abided with the AEC and NRC regulations and standards.

15-3. Nuclear Engineering Team Qualification

It is imperative that the engineering team to be very well familiar with the nuclear plant engineering requirement and all the regulations and standards issued by AEC and the NRC together with the project aspects regardless the power size of the plant required capacity. Both listed atomic regularity agency are very qualified by the local government.

The qualified engineer must be able to prepare all official reports, which is discussed and approved by the qualified engineer during several meetings with the listed agency. This requires several meetings between the design engineer/ engineers with nuclear authority to either prove the design document or not based on the authority final review of the submitted document. As listed earlier, the design engineer/ engineers must prepare all listed above documents to submit to several committees for approval by the authority. This also include preparing the PSAR and FSAR.

It is very possible that the engineering firm may be assigned to do several other requirements, such as preparing documents related to construction licensing and other requirements as the commission sees it necessary, and therefore the engineering firm should prove it's self-worthy to do any further assignment the owner may ask the firm to do.

16. Nuclear Regulations and Standards

The advanced educated world had restricted the work of nuclear energy to be used within determined applied regulations and standards based on scientific knowledge and professional experience for the benefits and safe application to warrant continued life for humans, animals, and plants around the world.

The Nuclear Regularity Commission (NRC) in the United States has published and enforced the application of many regulations and standards to guide the application of such standards when it comes to use nuclear energy within the safe boundary application and named the code, Code of Federal Regulations, Title 10, and other countries adopted such code application.

The NRC has confirmed the use of many standards and criteria application when it comes to the safe application of the nuclear energy. The following is a list of some those regulations and standards:

16-1. Commission Atomic Energy Regulations

16-2. Technical codes standards, such as IEEE standards; ASME standards; ANSI standards; IPCEA standards; and UL, DEMA, NEMA standards.

It is advised to review all pertinent regulatory standard details to understand the facts that application of those standards are founded for the safety of all.

CHAPTER 3

Electrical Engineering Study for Other Industrial Facilities

I. Nonpetrochemical Plant
II. Petrochemical Plant

I. Nonpetrochemical Plant Electrical Engineering Study

Most industrial facilities use contracted power source supply furnished by utility companies to power the operation of their equipment. This is done in accordance with legal contracts signed between the owner of the industrial facility called consumer, and the electrical utility company called the owner. However, the consumer may have a partial power generation plant to generate additional power if so desired, and such arrangement is agreed upon by the two sides in the signed contract.

The economic conditions play a major role to decide upon using the described arrangement or not. The industrial facility has the final decision in such case; however, the utility company encourages the consumer to buy electric energy from it by giving the industrial facility owners certain privileges included in a contractual agreement between both sides. Furthermore, the utility company makes it available to the industrial consumer that energy cost is proportional to the quantity of energy consumption within a specified range, and use a consumption factor as indicated in the contractual agreement, which can be renewed at the expiration of the signed contract. This privilege gives the industrial consumer a lower cost of energy should the energy consumption at the industrial plant go higher.

The industrial facility may have an option just in case the power load factor expiration date was not monitored, and as such the facility owner may wish to use some devices called load shedding devices to drop the excess load to below certain level (contracted level). The dropped load is generally considered not important to keep the facility going. This process is done to avoid put the facility power consumption within the contracted values.

It is the engineering firm who prepares the design of power distribution in the industrial facility and can recommend the design of what is called capacitors application to work with the large motor load to improve what is called motor power factor and hence reduce motor power demand a process which reflects a saving in the total power consumption when it comes to cost saving.

The method of correcting motor power factor is called power factor correction, and it has been proved that the cost of power factor correction devices pays for itself when it comes to reduce excess power saved by the described method, namely power factor correction.

The owner of an industrial facility presents to the utility company a preliminary study of power source requirement for a specified electric load delivered at the premises of the facility, which is a task in general a hired engineering company may do that. The engineering company may specify a requirement for a power transformer to step down the nearby power transmission potential level to a suitable level, which can be used to power the electrical equipment at the specified facility. The requirement of using a power transform has two folds. The power transmission lines distribution may have a potential level higher than the potential level of the load to be connected to, or the nearby transmission lines potential level is adequate to use directly at the premises, and other facility load can be provided with a lower potential level through a separate power transformer. In any event a hired engineering company may be required to prepare a study, be reviewed, and approved by the owner of the industrial facility, then later submitted to the utility company to review and to propose an adequate power service.

1. Industrial Plant Main Power Supply Engineering Study

Power source delivered to an industrial facility differs from one facility to another in accordance with the engineering study furnished to the particular facility, and sometimes the utility company may render its recommendation in that sense; however, there are several types of power source available to the consumer to choose from.

It is imperative first to select the location of the power source, and this is done with the cooperation of the utility company who can locate the nearest power transmission lines to the industrial facility location, a matter which is important for both sides when cost and potential level of the new facility to be constructed soon.

The utility company may offer several choices of power source schemes. The listed below schemes are mostly used:

- Loop Power Source Distribution Scheme
- Series Power Source Distribution Scheme

1-1. Loop Power Source Supply Scheme

The loop power source supply scheme is based on to connect the industrial facility with two lines. The first line is to connect the main power distribution board to the utility power transmission lines, and the second line is distributed from the main board to other power distribution panels in the industrial plant as shown in the following typical figures.

The looped power source supply lines scheme is considered the safe scheme among other schemes, which is sometimes an expensive scheme, but it provides a continuous power source supply in powering industrial facilities.

The power source loop scheme is usually connected to the main power supply panel through the main power transformer. The distribution of the power source is run from power supply panel to other panels, and so on.

1-2. Series Power Source Distribution Scheme

This type of power source supply depends on providing or connecting two or one power line connected between the utility power transmission lines and the industrial facility as may be recommended by the engineering study. The power source supply will be connected to local power distribution panels, where such power supply will be used to power local electric loads. The engineering study may recommend to connect such power source to an overhead bus bars ducts in addition to local power supply panels as dictated by the engineering study. As indicated earlier the power source is in most cases delivered by a main power transformer located between the utility company power transmission lines, and a suitable location near the industrial facility premises then connected to several power supply panels. The figure is shown above with other schemes.

2.   Local Electric Load Power Supply Source

Power supply source depends mostly on the size of the motor load. For motors which have rated capacity that exceeds 3,000 horsepower, the power supply source is rated generally not be less than 6,000 volt or 13,000 volts in accordance to what is generally available, and used at the industrial facility for such motor loads. For motor loads that are rated 500–2,000 horsepower, the potential level requirement should not be less than 4,000 volts. All motor loads that are rated below 500 horsepower, the supply potential should not be less than 480 volts. Most motor loads rated less than 500 horsepower can be supplied from power source supply equipment, which can be standing motor control centers or bus bars ducts to be specified by the engineering study. Usually motor loads that is half a horsepower or less, or lighting load can be power supplied from power supply panels rated 400/220/120 volts (three lines + neutral system).

Power distribution throughout industrial panels, and it will not differ whether the main supply in the facility uses loop distribution or series distribution power distribution scheme.

3. Industrial Plant Electrical Equipment

Industrial facility usually has several types of main electric equipment, which are used to supply power to its associated electric loads and are listed in sequence starting with power transmission lines as listed below:

3-1. Main circuit switcher, which is mounted on its associated steel pedestal to connect the facility main power transformer to the power transmission lines. The circuit switcher also protects the main power transformer through associated power fuses.

3-2. Facility main power transformer, which is connected to the main switch board on the secondary side of the transformer.

3-3. Main power switch board or main power switch-gear.

3-4. Local power supply panels or motor control centers or bus bars ducts, which are receiving power from the main power switch board through associated main protective control system such as main circuit breakers, or main power fuses in addition to power distribution molded case circuit breakers, and motors contactors with associated push buttons to operate the motors load. Should motors load be away from motors control centers (more than 50 feet), operating motors should have associated (disconnect switches) for safe maintaining the motors loads, which is a required standards application. At the end of this section. This should be applied to all local power supply panels listed in this book.

3-5. Lighting panels, which they may be in the form of three power lines + neutral, and associated distribution lighting transformer located close to the main lighting panels.

3-6. Direct current power panels in addition to battery-banks system which is charged by associated charger connected to the alternate current (AC) power source.

3-7. Battery-banks system located in an explosion proof room together with associate DC power panel, and battery banks plus associated equipment.

3-8. Emergency AC power supply system and associated converter. The facility may also use an emergency diesel generator or gas-turbine-driven generation unit or other type of power generation unit as suggested by the engineering company, and approved by the owner of the industrial facility.

The emergency facilities should also have an associated power distribution panel fully equipped with air circuit breakers and other protective and control equipment.

3-9. Telephone, public address system and instrumentation control system.

3-10. Emergency lighting and emergency exit signs system.

3-11. Fire protection system.

3-12. Required process control instrumentation and annunciator system, together with data logger system managing process control procedure.

3-13. Main control board.

3-14. Main control room.

The detailed engineering study will cover the requirement in every detail of the above-numerated items in an engineering book, which is associated with this write-up as soon as it is prepared and be printed.

Hazard Area Classification

II. Petrochemical Plant Engineering Study

Most petrochemical plants depend on electric power source, which is purchased from electric utility companies in accordance with legal contracts signed between the two sides; however, the petrochemical plants may generate partial power source on its own at its premises.

The electrical engineering study for the petrochemical plant is not entirely different from the one prepared for industrial facility except the fact that most electrical equipment used in the petrochemical plant have to meet the explosion-proof equipment local codes requirements in order to avoid the possibility of any explosion, or fire accidents due possibly to the presence of burning gas or liquid that may react with electrical equipment operation, and consequently may start an explosion or fire in that particular area.

As stated earlier in the engineering study of industrial facility the economic evaluation plays a major role in choosing between buying electric energy or to generate local electric power at the plant; however, sometimes the owner of the facility may combine both, and petrochemical facility is not any different. This is applicable when the facility contract to purchase electric energy from the utility company during frequent periods to be renewable between the contracted parties. This also has been discussed earlier in the engineering study of the industrial facility.

The owner of the new petrochemical facility will present to the utility company a preliminary engineering study prepared by a hired engineering company for the utility company to study, and to prepare an electrical power service to suit the requirement of the new petrochemical plant. The engineering company may also present a power service saving plans such as the use of power factor correction, and other services that the owner of the facility may see it fit such as the use of explosion-proof equipment and associated housing and control.

Technical and scientific standards application are set to identify the critical location of areas called hazardous areas shown below to be classified as designated below:

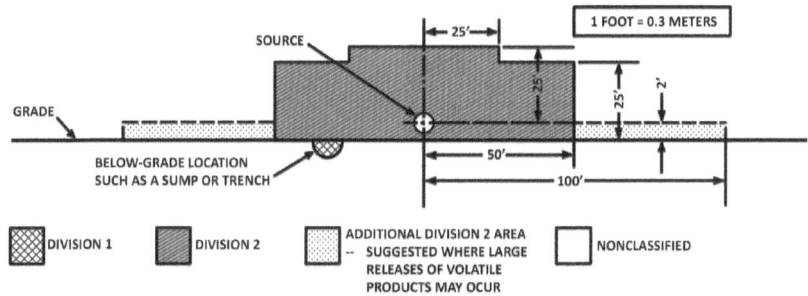

Note: Distances given are for typical refinery installations; they must be used with judgment, with consideration given to all factors discussed in the text. In some instances, greater or lesser distances may be justified.

Figure 1 - Adequately Ventilated Process Area with Heavier-Than-Air Gas Source Located Above Grade

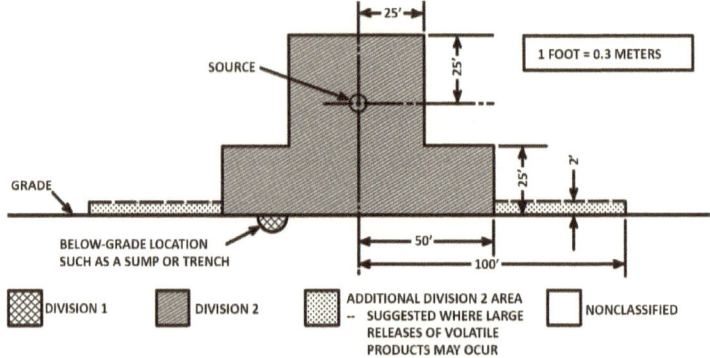

Note: Distances given are for typical refinery installations; they must be used with judgment, with consideration given to all factors discussed in the text. In some instances, greater or lesser distances may be justified.

Figure 2 - Adequately Ventilated Process Area with Heavier-Than-Air Gas Source Located Above Grade

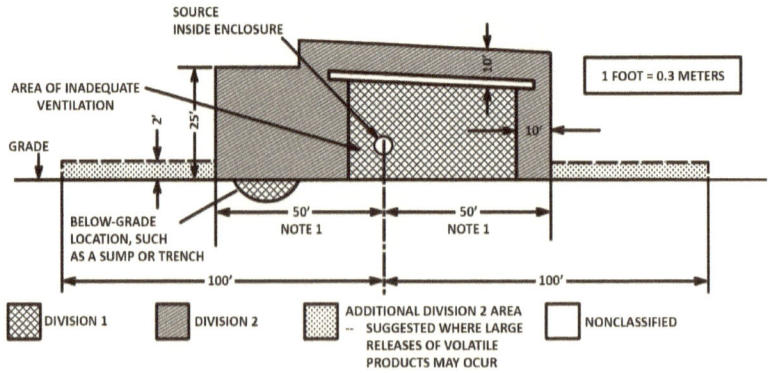

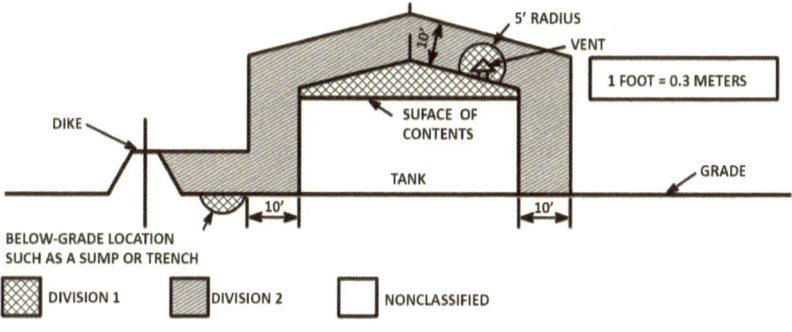

Notes:
1. For floating-roof tanks, the area above the tank roof and within the shell is classified Division 1.
2. High filling rates or blending operations involving Class I liquids may require extending the bounderies of classified areas.
3. Distances given are for typical refinery installations; they must be used with judgment, with consideration given to all factors discussed in the text.

Figure 4 - Refinery Tank with Heavier-Than-Air Gas Source

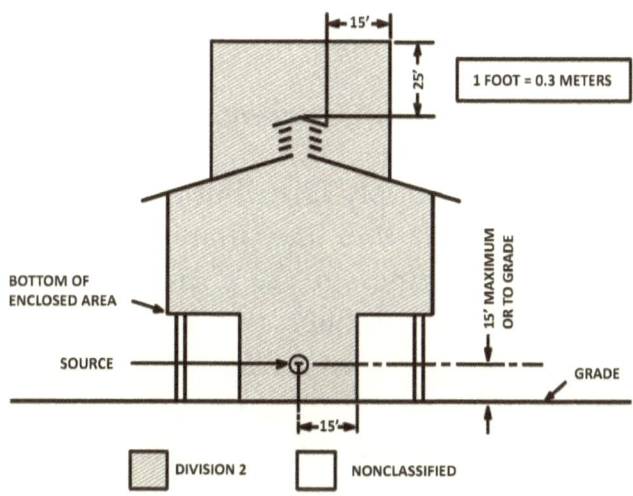

Note: Distances given are for typical refinery installations; they must be used with judgmnet, with consideration given to all factors discussed in the text.

Figure 5 - Adequately Ventilated Compressor Shelter with Lighter-Than-Air Gas Source

1. Petrochemical Plant Main Power Supply Engineering Study

The engineering study of the main power source supply may be different from any power source, and both the utility company and the engineering company may play a major role in determining what power source the petrochemical may want to use for the plant.

As stated earlier, the engineering company may select the exact location of power source supply, which is to be closer to the site of the new petrochemical plant with the help of the utility company who will provide the best location of the power source to connect the new plant main power to, and be it the nearest power transmission lines of the utility company. The utility company may offer the owner of the petrochemical plant several power supply schemes as listed below:

1-1. Loop Power Supply Distribution

1-2. Series Power Supply Distribution

- Plant Loop Power Supply Distribution

This scheme is to run two lines from the utility power company transmission lines to the chemical plant main power distribution board through two independent power transformers to connect the two sides of the main board, then run feeders from the plant main board to other power supply panels throughout the chemical plant in accordance with the engineering study made especially for the plant. The two sections of the main board are tied by a tie breaker in order to power supply the entire plant by a loop type of power supply distribution scheme. The plant local motor control centers and other power supply panels then are supplied with power by the described power distribution scheme. The power distribution schemes are shown above and similar to industrial plant except it should in compliance with hazardous areas location requirement.

The loop power distribution scheme is considered one of the safe scheme to use because of the two main power distribution lines supplied to the main control board in the petrochemical plant facility and can economically save to install local power generator based on the engineering study.

- Plant Main Series Power Supply Distribution

Petrochemical plant uses one main or two main power supply connections provided by the main power transmission lines, which will be connected to the plant main power distribution board in accordance with the engineering study, and the owner request.

This power source is distributed to other plant local power supply panels, which are located in different areas in the plant, and therefore supplying power to the electric motor loads as indicated by the design drawings.

- Plant Local Power Supply Panels

This type of power supply depends on the type of motors load and other type of mechanical load power supply requirements. Motors rated 3,000 HP or more are usually connected to local switch gears rated 6,000–15,000 volts in accordance with the engineering study. Motors load rated 500–2,000 HP are usually connected to local switch gears rated not less 4,000 volts. Motor loads rated below 500 HP is usually connected to motor control centers rated not less than 380 volts.

Motor loads rated 1/2 HP or less, and all lighting equipment are supplied by power supply panels rated 400/220 or 480/120 volts, three phases four wires as dictated by the engineering study.

The above-indicated power distribution scheme is very similar to the power distribution scheme in most cases is used in the series power distribution scheme of the industrial facility.

- Petrochemical Plant Electric Equipment

The plant uses several types of electric equipment, and it will be listed as indicated below starting with the main equipment:

- Plant main power transformer, which is connected on its high-voltage side with the utility power transmission lines through associated power circuit switcher, a device is used to connect and to protect the plant main power transformer.
- Main power switch board or main power switch gear, which is supplied with power through its main fused disconnect switch or main power circuit breaker located inside of the main switch gear enclosure.
- Power supply panels or motor control centers to power its associated motors load and other electric load in accordance with the engineering study.
- Lighting power supply panels for powering several local lighting equipment throughout the plant, and other lighting devices.
- Direct current power supply panels.
- Emergency power supply panels.
- Direct current power distribution panels.
- Battery banks and associated battery-bank Charger and associated equipment.
- Telephone and public address system and its associated equipment.
- Emergency lighting and exit signs.

- Main control room and main control board and associated annunciator system and data logger.
- Plant fire protection system.

4-1.Plant Main Power Transformer

The selection of the plant main power transformer depends on the power transmission lines' potential level, which is located near the plant facility, and if the potential level is above 600 volts, then there will be a need to use main power transformer. The utility company may contract the use and install such transformer with the owner of the facility, or the owner may purchase the required power transformer. In any event the required power transfer may be included in the engineering study together with the required instrument protection, such as the listed power circuit switcher and associated protective relays, and may not be the same size as required for power generation plant power transformer. (Refer to page 104 addressing other requirements).

The figure shown below are typical arrangement for requirement of power.

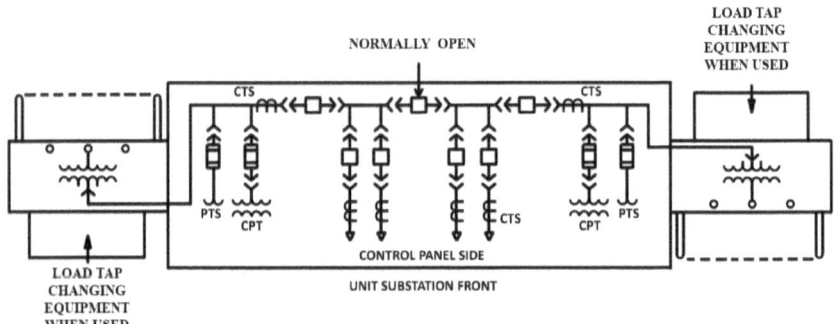

Fig. 4

Typical Secondary-selective-type Substation

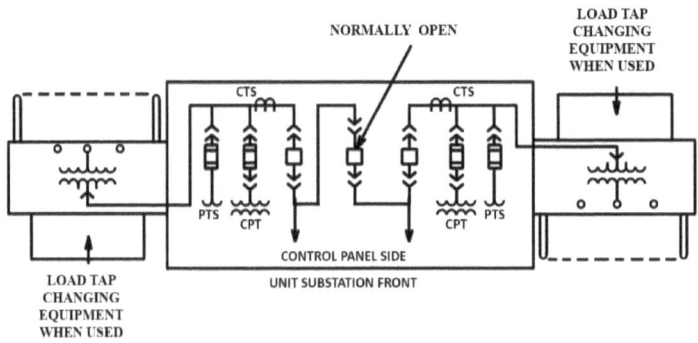

Fig. 5

4-2. Plant Main Power Switch Gear or Main Board

This power switch gear is similar to any switch gear as described earlier, which contains the main switch gear main circuit breaker, and branch circuit breakers, which may power other local supply panels throughout the plant as described in the engineering study.

4-3. Plant Local Power Supply Panels or Motor Control Centers

These power supply panels or motor control centers are distributed throughout the plant to be close to the motors load, or other loads, and may contain molded case circuit breakers or fused circuit switches, and associated control contactors in addition to remote disconnect switches for the safe power cut out the load during servicing (motor contactor and motor load.

It should be emphasized here that should the above-described equipment fall within a possible explosion area, then all equipment should meet the explosion-proof standards equipment.

As indicated earlier the use of what is called power factor correction for induction motor types is recommended, and the petrochemical facility may adopt such feature for the sake of avoiding power losses.

Motor control centers are used in the petrochemical plants, which are special cubical containing the horizontal and vertical electric bars and associated air circuit breakers installed inside the cubicles in a safe manners. The surface of such cubicles is equipped with control lights, control switches, and also control instrumentation as required by the engineering study.

Petrochemical facilities are considered usually limited industrial facilities, which may experience exposure of gas and liquid explosive and burning materials; therefore, the facilities are subject to adhere to conservative standards application such as the use of explosion-proof materials. Since the facilities are described to be limited-area industrial facilities, back-to-back cubicles assembly type of motor control centers equipment is recommended to be used for powering motor loads, and other loads.

4-4. Plant Lighting Panels

Petrochemical plants are recommended to use special types of lighting requirement, which is usually power supplied by light panels installed together with associated lighting transformer at the motor control center throughout the plants.

4-5. Plant Direct Current (DC) Power Supply

Petrochemical plants usually require a DC power source and associated battery banks and battery chargers supplied by an AC power source. A required DC generator can be substituted in lieu of battery banks to power a certain kind of DC load. The DC power source will be distributed through associated DC power supply panels, which include the main circuit breaker and distribution circuit breakers. The entire DC power source will be located in a separate room, which should meet the explosion-proof equipment requirements.

4-6. Plant Emergency Power Supply Equipment

The emergency power system is one of the essential equipment to maintain power and consequently to keep up with product quality control requirement. The emergency system is the type of diesel generator unit or the type of gas-turbine unit, with all auxiliary equipment such as control panel and power distribution panel, together with fuel tank, which is related to the type of the emergency unit (diesel fuel or gas fuel) and other associated (main fuel tank and daily fuel tank). The generated power will be the three-phase, four-wires type with the required potential level as specified in the engineering study.

4-7. Plant Telephone and Public Address System

This requirement is similar to other plants equipment when it comes to plan a telephone distribution system throughout the plant including the outside telephone service, and which include the requirement of public address system, which include loud speakers and other equipment. The engineering study will address all of the requirement, including any control instrumentation, such as process automatic control dictated by the instrument engineer.

4-8. Plant Fire Protection System

Plant fire protection system is a demanded required regulation for the safety of the plant employees as well as for the plant safe equipment operation, which all insurance companies require to have when it comes to financial compensation. In the event the petrochemical plant has any kind of hazardous accidents. It has been determined by the Insurance companies that no financial compensation for any claim if it was found that the plant did not have adequate protection to prevent accidents that may occur at the plant facility.

4-9. Plant Emergency Lighting and Exit Signs

Emergency lighting and exit signs are located in most of the closed areas of the plant to help for the safe exit of people found in the plant during fire accidents and other emergency accidents and consider a safe procedure applied in most operating industrial plants.

4-10. Plant Main Control Room

As it has been stated, the control room is the brain operating room in any facility; petrochemical plants are not any different. The control room will house a main control board, which has several control instrumentations, process sequence mimic bus, annunciator system, data logger, and several control switches, and alarm point.

BOOK CLOSING COMMENT

The engineering study regardless what it may include at several stages of the study; it may present some details, however. The final detailed engineering study should have complete engineering study to help constructing a safe facility, calling it power generation plants (nuclear as well conventional type) or industrial facilities (call it chemical or conventional facility). The detailed engineering study will be covered in the second engineering book to be soon published. Please look forward. I thank you.

Dr. Mohamed S. Khorsheed

To the Publisher:
All Content of this book is certified to be the author's own details, which are lawfully preserved. Any copy of the content is a lawful violation of the preserved rights of the author.

www.ingramcontent.com/pod-product-compliance
Lightning Source LLC
Chambersburg PA
CBHW030755180526
45163CB00003B/1027